WATER RESOURCE PLANNING, DEVELOPMENT AND MANAGEMENT

EPA PROTECTION OF DRINKING WATER FROM INJECTION OF FLUIDS USED IN OIL AND GAS PRODUCTION

WATER RESOURCE PLANNING, DEVELOPMENT AND MANAGEMENT

Additional books in this series can be found on Nova's website under the Series tab.

Additional e-books in this series can be found on Nova's website under the e-book tab.

WATER RESOURCE PLANNING, DEVELOPMENT AND MANAGEMENT

EPA PROTECTION OF DRINKING WATER FROM INJECTION OF FLUIDS USED IN OIL AND GAS PRODUCTION

COURTNEY D. WALSH
EDITOR

nova publishers
New York

Library of Congress Cataloging-in-Publication Data

ISBN: 978-1-63463-535-6

Published by Nova Science Publishers, Inc. † New York

CONTENTS

PREFACE

This book examines EPA and state roles, responsibilities, and resources for the program; safeguards to protect drinking water; EPA oversight and enforcement of class II programs; and the reliability of program data for reporting. GAO reviewed federal and state laws and regulations.

Chapter 1 – Every day in the United States, at least 2 billion gallons of fluids are injected into over 172,000 wells to enhance oil and gas production, or to dispose of fluids brought to the surface during the extraction of oil and gas resources. These wells are subject to regulation to protect drinking water sources under EPA's UIC class II program and approved state class II programs. Because much of the population relies on underground sources for drinking water, these wells have raised concerns about the safety of the nation's drinking water.

GAO was asked to review EPA's oversight of the class II program. This report examines (1) EPA and state roles, responsibilities, and resources for the program, (2) safeguards to protect drinking water, (3) EPA oversight and enforcement of class II programs, and (4) the reliability of program data for reporting. GAO reviewed federal and state laws and regulations. GAO interviewed EPA and state officials and reviewed class II programs from a nongeneralizable sample of eight states selected on the basis of shale oil and gas regions and the highest number of class II wells.

Chapter 2 – This is a report by U.S. Government Accountability Office on Drinking Water: Characterization of Injected Fluids Associated with Oil and Gas Production, dated September 23, 2014.

Chapter 3 – This document is intended to summarize the basis and purpose underlying the underground injection control regulations promulgated in 40 CFR Part 146. It sets forth generally the reasoning behind the Agency's regulatory choices and references data upon which EPA relied.

This statement first discusses the categorization of wells. It then surveys the major pathways which contaminants can take to enter underground sources of drinking water and the requirements which these regulations impose to assure that movement of fluids resulting from well injection does not contaminate underground sources of drinking water. Programmatic requirements of Part 146, such as monitoring and reporting, are covered in the concluding section.

In: EPA Protection of Drinking Water
Editor: Courtney D. Walsh
ISBN: 978-1-63463-535-6

Chapter 1

DRINKING WATER: EPA PROGRAM TO PROTECT UNDERGROUND SOURCES FROM INJECTION OF FLUIDS ASSOCIATED WITH OIL AND GAS PRODUCTION NEEDS IMPROVEMENT*

United States Government Accountability Office

WHY GAO DID THIS STUDY

Every day in the United States, at least 2 billion gallons of fluids are injected into over 172,000 wells to enhance oil and gas production, or to dispose of fluids brought to the surface during the extraction of oil and gas resources. These wells are subject to regulation to protect drinking water sources under EPA's UIC class II program and approved state class II programs. Because much of the population relies on underground sources for drinking water, these wells have raised concerns about the safety of the nation's drinking water.

GAO was asked to review EPA's oversight of the class II program. This report examines (1) EPA and state roles, responsibilities, and resources for the

* This is an edited, reformatted and augmented version of the United States Government Accountability Office publication, GAO-14-555, dated June 2014.

program, (2) safeguards to protect drinking water, (3) EPA oversight and enforcement of class II programs, and (4) the reliability of program data for reporting. GAO reviewed federal and state laws and regulations. GAO interviewed EPA and state officials and reviewed class II programs from a nongeneralizable sample of eight states selected on the basis of shale oil and gas regions and the highest number of class II wells.

What GAO Recommends

GAO recommends that, among other things, EPA review emerging risks related to class II program safeguards and ensure that it can effectively oversee and efficiently enforce class II programs. EPA agreed with all but the enforcement recommendation. GAO continues to believe that EPA should take actions to ensure it can enforce state class II regulations, as discussed in the report.

What GAO Found

The Environmental Protection Agency's (EPA) role in the Underground Injection Control (UIC) class II program is to oversee and enforce fluid injection into wells associated with oil and gas production, known as class II wells. EPA has approved 39 states to manage their own class II programs, and EPA regions are responsible for managing the programs in remaining states. EPA regions and states use a mix of resources to manage class II programs, including EPA grant funding, state funding, and federal and state personnel. EPA's UIC grant funding has remained at about $11 million for at least the past 10 years.

Class II programs from the eight selected states that GAO reviewed have safeguards, such as construction requirements for injection wells, to protect against contamination of underground sources of drinking water. Programs in two states are managed by EPA and rely on EPA safeguards, while the remaining six programs are state managed and have their own safeguards that EPA deemed effective at preventing such contamination. Overall, EPA and state program officials reported that these safeguards are protective, resulting in few known incidents of contamination. However, the safeguards do not address emerging underground injection risks, such as seismic activity and overly high pressure in geologic formations leading to surface outbreaks of fluids. EPA officials said they manage these risks on a state-by-state basis, and some states have additional safeguards to address them. EPA has tasked its UIC Technical Workgroup with reviewing induced seismicity associated with injection wells and possible safeguards, but it does not plan reviews of other emerging risks, such as high pressure in formations. Without reviews of these risks, class II programs may not have the information necessary to fully protect underground drinking water.

EPA is not consistently conducting two key oversight and enforcement activities for class II programs. First, EPA does not consistently conduct annual on-site state program evaluations as directed in guidance because, according to some EPA officials, the agency does not have the resources to do so. The agency has not, however, evaluated its guidance, which dates from the 1980s, to determine which activities are essential for effective oversight. Without such an evaluation, EPA does not know what oversight activities are most effective or necessary. Second, to enforce state class II requirements, under current agency regulations, EPA must approve and incorporate state program requirements and any changes to them into federal regulations through a rulemaking. EPA has not incorporated all such requirements and changes into federal regulations and, as a result, may not be able to enforce all state program requirements. Some EPA officials said that incorporating changes into federal regulations through the rulemaking process is burdensome and time-consuming. EPA has not, however, evaluated alternatives for a more efficient process to approve and incorporate state program requirements and changes into regulations. Without incorporating these requirements and changes into federal regulations, EPA cannot enforce them if a state does not take action or requests EPA's assistance to take action.

EPA collects a large amount of data on each class II program, but the data are not reliable (i.e., complete or comparable) to report at a national level. EPA is working on a national database that will allow it to report UIC results at a national level, but the database will not be fully implemented for at least 2 to 3 years.

ABBREVIATIONS

SDWA	Safe Drinking Water Act
AOR	area of review
EPA	Environmental Protection Agency

mg/l	milligrams per liter
UIC	Underground Injection Control

***

June 27, 2014

Congressional Requesters

Every day in the United States, at least 2 billion gallons of fluids are injected into underground formations to enhance oil and gas production or to dispose of fluids brought to the surface during the extraction of oil and gas resources. These fluids consist largely of saltwater and may contain pollutants such as chlorides, hydrocarbons, and naturally occurring radioactive materials. Using injection wells that rely on gravity or pressure, the fluids are injected deep underground into porous rock formations, such as sandstone, that are typically below aquifers that can, or do, supply drinking water. Because a significant percentage of the population gets its drinking water from underground aquifers, these wells have raised concerns about the safety of the nation's drinking water.

Wells used for injecting fluids associated with the extraction of oil and gas resources are known as class II injection wells.[1] As of 2012, there were over 170,000 class II injection wells in the United States, located in states as diverse as California, New Mexico, Oklahoma, Pennsylvania, Texas, and Virginia. To protect underground sources of drinking water from contamination, class II injection wells are subject to regulation by the Underground Injection Control (UIC) program overseen by the Environmental Protection Agency (EPA) under the Safe Drinking Water Act (act), which is managed by state regulatory agencies and, in some cases, EPA.[2] Under the UIC program, EPA and states rely on safeguards to prevent fluids from leaking into aquifers that can be used as underground sources of drinking water. These safeguards require well operators to among other things, meet technical standards for constructing, operating, testing, and monitoring injection wells.

The UIC program regulates three types of class II wells associated with oil and gas production: (1) enhanced recovery wells into which brine, water, steam, carbon dioxide, or other fluids and gases are injected into oil- or gas-bearing formations to increase the recovery of residual oil and gas; (2) disposal wells into which brines and other fluids brought to the surface during oil and gas production activities are injected to dispose of them; and (3) storage wells into which liquid petroleum products are injected, generally as part of the U.S. Strategic Petroleum Reserve.[3] EPA estimates that approximately 80 percent of class II wells are enhanced recovery wells.

States request approval from EPA to manage the UIC programs in their respective borders, including the class II programs. Under the act, the EPA Administrator approves state programs for one or more classes of wells through a rulemaking process, with public notice and comment, and EPA updates federal regulations to reflect the approved program. Once EPA has approved a state's program, the state has primary management and enforcement responsibility for its UIC program, known as primacy. In states that do not have approval to manage their programs, EPA regional offices manage the programs in the state directly. Twenty-five states with class II wells manage their class II programs and regulate over 95 percent of the class II wells nationwide. Eight states with class II wells have programs managed by EPA regional offices. The remaining 17 states have no class II wells. (See app. I for a list of the states that manage their programs and the states in which EPA manages the program.)

Domestic production of oil and gas has increased dramatically in the last several years, with corresponding increases in the wastewater resulting from production processes.[4] As we reported in September 2012, oil and gas production from shale formations increased more than 4-fold in recent years and is expected to increase in the future. Improvements in technology have made it economically feasible to extract oil and natural gas from unconventional sources that were not previously profitable. Specifically, hydraulic fracturing combined with horizontal drilling has increased domestic production from unconventional sources found in shale, tight sandstone, and coalbed methane formations. Fracturing processes can use, per well, between approximately 2 million and 6 million gallons of water combined with sand and chemical additives. While hydraulic fracturing involves the injection of fluids underground, the Energy Policy Act of 2005 exempted the process of injecting fluids—other than diesel fuel—into a well to hydraulically fracture formations.[5] However, water that is produced from formations during oil and gas production, including water from hydraulic fracturing activities that flows back out of the well, needs to be disposed of or reused. Water that is injected underground for disposal or to enhance recovery is regulated under the class II UIC program.

The increase in wastewater associated with oil and gas production—both produced water and flowback water—has created a need for additional class II wells. The number of class II wells grew from approximately 144,000 in 2005 to over 172,000 in fiscal year 2012.[6] Concerns have been raised about the potential environmental and public health effects associated with the large amounts of wastewater generated during and after the fracturing process. In 2012, we reported that produced water and fracturing fluids contain a wide range of contaminants that pose a risk to groundwater quality, if not properly managed. At high levels, exposure to some of the contaminants in produced water could adversely affect human health and the environment. Similarly, some additives used in fracturing fluids, such as diesel fuel, are known to have toxic constituents. We also reported that the extent and severity of environmental and public health risks associated with shale oil and gas development depend in part upon federal and state regulations and oversight, including the UIC program.[7]

Given the increased wastewater disposal needs associated with increasing oil and gas production, you requested that we provide information on EPA's oversight of the class II program. In this report, we examine: (1) EPA and state roles, responsibilities, and resources in managing the class II program; (2) EPA and selected state safeguards to protect underground sources of drinking water; (3) EPA's oversight and enforcement of class II programs; and (4) the reliability of data to report on the class II program nationwide.

To perform this work, we identified and analyzed relevant legislation, regulations, guidance, and program documentation, as well as documentation on oversight and data management. We also interviewed officials from EPA at headquarters and eight regional offices, as well as state officials who manage the class II program in their state. Further, we selected a nonprobability sample of eight states with class II programs— California, Colorado, Kentucky, North Dakota, Ohio, Oklahoma, Pennsylvania, and Texas—to analyze the class II program at the state level. The results of our work in these states cannot be generalized to other states, but do provide detailed examples of EPA's and states' management of class II programs. We selected these states from the six shale oil and gas regions as defined by the Energy Information Administration. For each of the six shale regions, we selected at least one state that had among the highest number of class II injection wells, ensuring that we had both states that EPA had approved to manage their class II programs (state programs), such as California, Colorado, North Dakota, Ohio, Oklahoma and Texas, as well as states in which EPA regions manage the class II program (EPA-managed programs), such as Kentucky and Pennsylvania.[8]

To address our first objective, we reviewed the act, which establishes the UIC program, and describes the process by which states may be approved to manage their own UIC programs. We interviewed EPA officials at headquarters and the eight regional offices with states that have oil and gas activity and class II underground injection control programs about their management of the program. We also spoke with officials in six of our eight selected states about the day-to-day management and operation of their class II programs, and we requested data on the relative share of the class II program budget contributed by the states. Similarly, we spoke to the regional office staff responsible for managing the programs in Kentucky and Pennsylvania about their management of the class II programs in these states.

To address our second objective, we identified and reviewed relevant federal regulations and program guidance that establish minimum federal safeguards to protect underground sources of drinking water from contamination by injected fluids. These federal regulations applied to two of our eight selected states—Kentucky and Pennsylvania. We also identified and reviewed state regulations and program guidance regarding safeguards in the six remaining states in our sample. To complement our analysis of law, regulations, and guidance, we interviewed officials in EPA headquarters and eight regional offices responsible for overseeing the UIC program in the selected states, and interviewed state and appropriate regional officials about the implementation of these safeguards in the eight selected states. Our review included a summary and comparison of the regulations and guidance that establish state and EPA-managed program safeguards, but we did not analyze the technical sufficiency of those safeguards.

To address our third objective, we reviewed EPA regulations and guidance on approving state programs and revisions, and EPA guidance that defines effective oversight of class II programs by EPA headquarters and regional offices. We also spoke with EPA headquarters and regional officials about how this guidance is implemented and with state and regional officials about the oversight of programs in the eight states we reviewed. We reviewed documentation from state and regional evaluations, annual reports submitted by states to EPA, and agreements between EPA and states outlining the terms of their UIC programs.

Finally, in support of the fourth objective, we obtained and evaluated class II program data from reporting forms that state and EPA-managed programs submit to EPA, and well inventory data collected by EPA in a Web-based system. To assess the reliability of the data from reporting forms, we interviewed EPA headquarters officials about their processes for collecting and managing the data and tested the data for completeness. We also interviewed EPA regional officials and state officials about their processes for completing and submitting the data and checked for consistency in how they reported the data. To assess the reliability of well inventory data, we interviewed EPA officials about their collection of the data and any limitations in using it. We determined that the inventory data were reliable for our reporting purposes, and that the data from reporting forms were reliable for purposes of reporting individual state data, but not for aggregating and reporting national data. In addition, we spoke to EPA officials regarding a national database that it is developing, which is not yet complete or used for reporting. We discussed its capabilities and purpose, as well as reviewing documentation on database development.

We conducted this performance audit from December 2012 to June 2014, in accordance with generally accepted government auditing standards. Those standards require that we plan and perform the audit to obtain sufficient, appropriate evidence to provide a reasonable basis for our findings and conclusions based on our audit objectives. We believe the evidence obtained provides a reasonable basis for our findings and conclusions based on our audit objectives. A more detailed description of our objectives, scope, and methodology is presented in appendix III.

BACKGROUND

This section presents information on the use of injection wells in the oil and gas industry and key provisions of the Safe Drinking Water Act and EPA regulations related to protecting underground sources of drinking water, and EPA UIC regulations and evaluations.

Oil and Gas Production and Use of Underground Injection

After a long period of declining production that lasted through 2005, annual U.S. production of oil and gas—and associated wastewater—has experienced significant growth. This increase is due in large part to development of oil and gas from unconventional formations. For example, shale oil production increased more than 5-fold, from 39 million to about 217 million barrels from 2007 through 2011, while shale gas production increased approximately 4-fold, from 1.6 to about 7.2 trillion cubic feet, over the same period. Oil and gas production, including production from unconventional oil and gas bearing formations, involves removing hydrocarbon bearing fluids from geologic formations underground. Conventional oil and natural gas are found in deep, porous rock or reservoirs and can flow under natural pressure to the surface after drilling. In contrast to the free-flowing resources found in conventional formations, the low permeability of some formations, including shale, means that oil and gas trapped in the formation cannot move easily from the rock.

As a result of geologic differences, the methods used to extract hydrocarbon resources, including the amounts of water used and wastewater produced during extraction, vary widely.[9] Some oil and gas can be developed by drilling a well and relying on the natural pressure in the formation to push the oil or gas to the surface. When an oil or gas producing reservoir is depleted—that is, no longer producing oil or gas—a producer can inject fluids, such as saltwater, into the reservoir to increase the pressure in the formation and move the residual oil and gas toward a well and the surface for collection, a process known as enhanced recovery (See figure 1.).[10]

Similarly, some hydrocarbons are produced through the use of hydraulic fracturing. Hydraulic fracturing involves the injection of liquid under pressure to fracture the rock in geologic formations such as shale to create fractures and allow hydrocarbons to flow more freely from the formation into the well for collection. The liquids used in this process consist primarily of water, but also include chemicals, and sand or other agents for holding open the fractures (proppants). When a well is no longer economically viable, a producer may decide to cease production from it. At some point after production stops, wells may be converted from production wells to disposal wells, where wastewater can be injected for disposal. If the well is not used for any of these purposes, it becomes idle or inactive and will eventually be plugged and abandoned.

The process of extracting oil and gas creates several waste streams that must be managed. Key among these waste streams is the water (brine) produced along with the oil or gas during production. Produced water may include water that occurs naturally in the formation, water or other liquids that were injected into the formation to enhance recovery during the drilling or production process, and flowback water, which consists of the water, chemicals, and proppants used for hydraulic fracturing (fracturing fluids). Over 90 percent of the water produced during oil and gas operations is disposed of underground, either through injection for disposal or for enhanced recovery. The remaining water may be discharged to surface water such as streams or lakes, stored in surface impoundments, reused for agricultural purposes such as irrigation or livestock, or reused for hydraulic fracturing.[11]

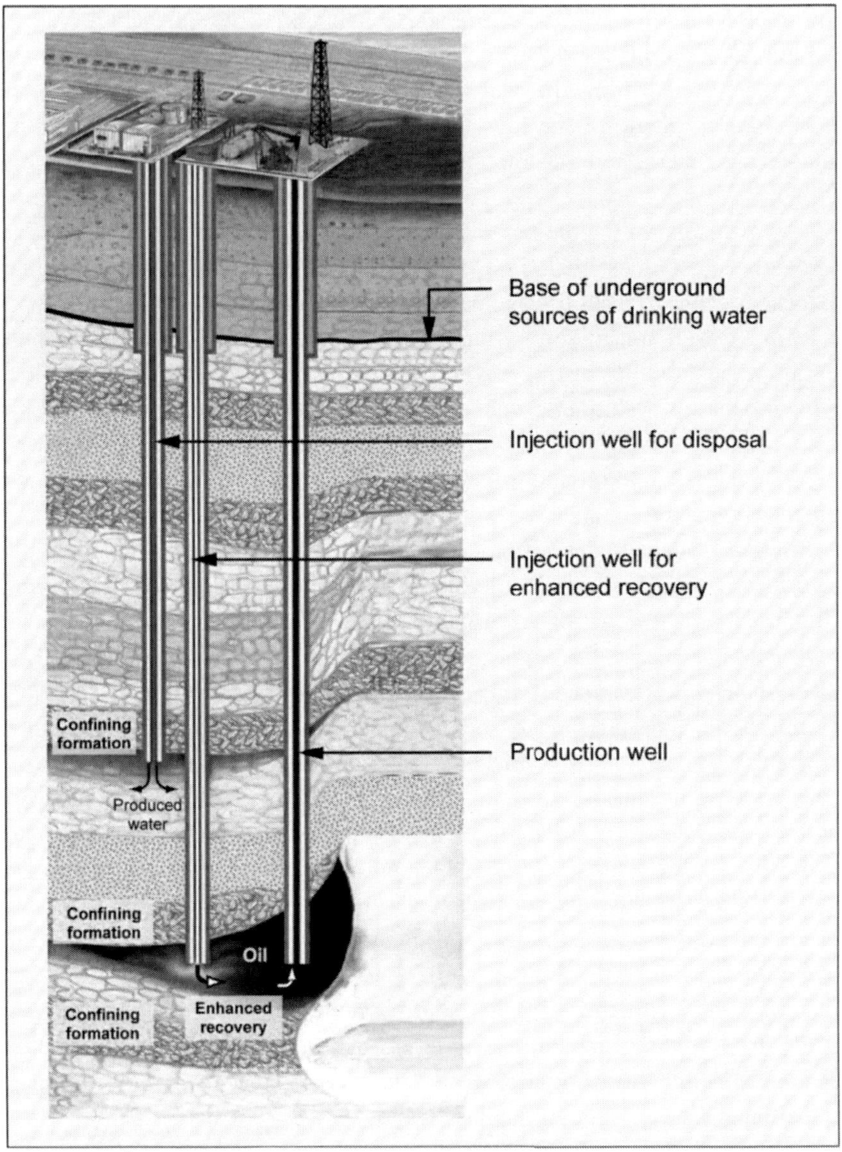

Source: EPA. | GAO-14-555.

Note: In 2005, the Energy Policy Act amended the Safe Drinking Water Act to exempt the underground injection of fluids associated with hydraulic fracturing operations related to oil, gas, or geothermal production activities from regulation under class II programs, except in cases where diesel fuels are used in the fracturing process.

Figure 1. Injection Wells for Enhanced Production and Wastewater Disposal.

Differences in geologic formations can affect the pressure and volume of wastewater injected underground for either disposal or reuse in enhancement of oil and gas production. Layers of sediment and rock deposited over time create different underground formations with different characteristics, depending on the material that was deposited and various geologic processes. These characteristics include how porous and permeable a rock is—both characteristics that determine how well a formation can accept and hold fluids such as wastewater. Generally, fluids are injected into porous, permeable reservoirs below an upper confining layer, or geologic formation with low permeability through which significant quantities of liquids cannot move. The confining layer serves as a barrier of protection between the reservoir where fluids are being injected, known as the injection zone, and underground sources of drinking water (see figure 1). Ideally, an injection zone is sealed above and below by continuous, impermeable rock formations and is large enough to keep injected fluids from reaching pressures great enough to fracture the confining rock layers.

The Safe Drinking Water Act and the UIC Program

Much of the nation relies on groundwater as a source of drinking water, and reflecting this fact, Congress included groundwater protection provisions in the 1974 Safe Drinking Water Act, as amended. The act, among other things, required EPA to establish the UIC program, including the oversight of state programs regulating underground injection activities, to prevent the endangerment of underground sources of drinking water. Before the act's passage, there was no overall federal regulation of injection activities, although some individual states regulated injection wells. Because of concerns about leaks from injection wells and the potential contamination of underground sources of drinking water, Congress established the national program in the Safe Drinking Water Act.

The Safe Drinking Water Act requires EPA to establish regulations for state programs with minimum requirements to prevent injection of fluids that endanger underground sources of drinking water. These minimum requirements are to prohibit the injection of fluids underground unless permitted and are to include, among other things, requirements for monitoring of and reporting on injection wells. In implementing the program, EPA has defined underground sources of drinking water as an aquifer, or part of an aquifer, that has not been exempted from regulation and either supplies a public water system or contains a sufficient quantity of water to supply a public water system and (1) currently supplies drinking water for human consumption or (2) which, among other things, contains fewer than 10,000 milligrams per liter (mg/l or parts per million) total dissolved solids.[12] Aquifers can be exempted from protection under the act and used for injection. Operators may apply for an exemption for a particular aquifer from EPA and, if granted, operators may inject fluids into the aquifer.

The act, as amended through 1977, placed three conditions on the UIC program. First, EPA's program regulations must not interfere with or impede the underground injection of brine or other fluids brought to the surface in connection with oil or natural gas production or natural gas storage operations, or any underground injection for the secondary or tertiary recovery of oil or natural gas, unless such requirements are essential to protecting underground sources of drinking water. Second, EPA's regulations of class II wells should provide for varying geological, hydrological, and historical conditions within and among the states. Third, EPA's program must not unnecessarily disrupt existing state UIC programs.

The act also included provisions for states to apply for and receive approval from EPA to manage one or more of the UIC programs in their state (each class of well is managed as a program).[13] Under the act, to gain EPA approval, a state generally must adopt and implement a program that meets the minimum requirements established under EPA regulations and conduct reporting as EPA requires. EPA must approve or disapprove a state's program through a rulemaking process, and EPA's practice is to enter into a memorandum of agreement with each state, outlining state and federal responsibilities for program management and oversight. States with approved programs—state programs—manage the permitting, inspection, and enforcement of injection wells in their states under EPA oversight.

Congress amended the act in 1980, creating an alternative process for states to receive approval for their UIC class II programs specifically.[14] Using this alternative, in lieu of adopting EPA's minimum requirements, a state can seek approval to manage its own program by demonstrating to EPA that its program is effective in preventing the contamination of underground sources of drinking water.[15] Similar to the conventional approval process, states are required to submit an application to EPA and enter into a memorandum of agreement with EPA laying out the components of their program. However, EPA has discretion when making the determination that the state's program is effective at protecting underground sources of drinking water. To help states seeking approval for their programs under this approach, EPA developed guidance outlining what would constitute a program that is effective in preventing contamination of underground sources of drinking water.

Most states have class II programs that have been approved by EPA under one or the other process, although some of these states do not have class II injection wells. EPA has approved 16 states under the conventional process (2 with class II injection wells), and 23 state programs under the alternative process (all 23 with class II injection wells), with most states approved in the 1980's soon after the program's inception (see app. I for a list of states). Eleven states do not have program approval (eight with class II injection wells), and the programs in these states are managed by six EPA regional offices.[16]

EPA UIC Regulations and Evaluations

In 1980, EPA issued regulations that established minimum requirements for state programs, and through the mid-1990s, issued guidance for the class II programs. Until recently, EPA has made few significant additions or updates to the regulations and guidance, which have remained in effect to govern EPA's class II program. In 1988, about a decade after EPA issued its regulations, the agency initiated a midcourse evaluation of the class II program to determine whether any changes were needed to improve the program.[17] EPA evaluated the adequacy of program regulations and safeguards, including various safeguards such as mechanical integrity tests, and found that most of the regulations and corresponding safeguards were adequate to protect underground sources of drinking water and needed no changes. EPA's evaluation also identified some areas of the program that could be improved, such as the temporary abandonment of injection wells and construction requirements for injection wells. In 1992, in a follow-on study, an EPA-chartered workgroup recommended that the agency amend its regulations to ensure that wells were abandoned in such a way that they would not provide a conduit for injected fluids.[18] The study also recommended that well construction requirements be changed so that all wells drilled after the regulations went into effect have multiple layers of steel casing and cement through formations that had a certain water quality. According to EPA officials, EPA developed draft regulations to address some of the recommendations from the evaluation and follow-on study, but the draft regulations were never published. EPA announced in 1995 that it would not pursue a change to regulations, and no subsequent action was taken.

In 1996, EPA created the UIC Technical Workgroup to provide a forum to evaluate ongoing technical issues in the UIC program. The workgroup has members from EPA, states, and environmental organizations, which discuss, review, and resolve technical matters related to underground injection. The workgroup's objectives include promoting consistent implementation of the UIC program and assisting with the development of regulatory revisions and technical guidance. Since its inception, the workgroup has reviewed a number of technical issues surrounding class II wells, including well construction standards and the types of fluids that are acceptable for injection into wells.

In 2005, the Energy Policy Act amended the Safe Drinking Water Act to exempt the underground injection of fluids associated with hydraulic fracturing operations related to oil, gas, or geothermal production activities from regulation under class II programs, except in cases where diesel fuels are used in the fracturing process.[19] In February 2014, EPA issued class II permitting guidance for hydraulic fracturing operations using diesel fuels. The guidance includes technical recommendations for EPA permit writers.

EPA Oversees Class II Programs, for Which EPA and States Provide a Mix of Resources

EPA and states have separate roles and responsibilities for class II programs and provide a mix of resources to support them. EPA headquarters oversees the class II program by setting program regulations and guidance, while regions focus their oversight on evaluating state programs and state reporting. State programs, as well as EPA-managed programs, are largely responsible for the day-to-day management of the class II program, including providing permits, conducting inspections, enforcing regulations and guidance, and collecting data and reporting. EPA and states provide a mix of resources to support the class II program including EPA grants, EPA and state personnel, and state funding.

EPA Oversees State and EPA-Managed Programs

EPA headquarters regulates and oversees state class II programs, in addition to all other UIC programs, by setting program regulations and guidance and providing grant funding to state programs. EPA's regional offices primarily oversee states that have approved UIC programs—particularly class II programs—by reviewing grants and conducting annual evaluations. EPA's primary responsibilities include the following:

- *Establishes program regulations and guidance.* EPA issues and, in some cases, updates minimum federal regulations and develops guidance, to assist with the implementation of UIC program requirements, including those established in the Safe Drinking Water Act. In addition, EPA develops guidance for state and EPA-managed programs on technical program requirements such as pressure testing and well construction.
- *Approves state program applications and state program revisions.* EPA reviews and approves state applications to manage their own programs. The act requires EPA to approve state programs, and certain revisions, by rule, meaning that EPA conducts a rulemaking including public notice and opportunity to comment. EPA has issued regulations and guidance concerning the process for approving state programs and revisions. Currently, 39 states have approved class II programs, and, according to EPA officials, there are two applications pending approval. See appendix I for a list of states with approved programs and appendix II for a description of the approval process.
- *Enforcement of state class II regulations.* Once state programs are approved, EPA incorporates state program regulations into federal regulations, which gives EPA the ability to enforce state program requirements in cases where the state does not take action or requests EPA assistance.[20] To incorporate state program requirements into federal regulations, EPA conducts a rulemaking process that provides public notice of the proposed regulations and the opportunity for any person or organization to review the requirements and submit comments in writing for a 30- to 90-day period. EPA generally provides a response to the significant issues raised during the comment period and discusses any changes made to the regulation as a result, and it publishes the text of the final regulation in the Federal Register. The Code of Federal Regulations provides a public record of approved state programs that can be enforced by EPA.
- *Oversees state programs.* EPA regions oversee state programs. EPA guidance issued in 1983 provides details of effective oversight of state

programs, including several steps that regions are to take. EPA regional officials are expected to conduct annual on-site evaluations of state managed programs that usually involve, among other things, an on-site meeting with state class II officials to discuss program performance and a review of permitting and inspection files, both of which are intended to help determine whether the state program is meeting the requirement to protect underground sources of drinking water from endangerment by underground injections. EPA headquarters is responsible for conducting the annual evaluation of EPA-managed programs.

- *Allocates and administers grants.* EPA allocates grant funding to state programs using a formula that includes the number of underground injection wells in each state, state population, and state land area. To receive funding, states are required to submit a grant application with a work plan that estimates the amount of federal funds and state funds needed during the state's fiscal year, and the number of personnel the funding will support to accomplish specific activities such as permitting, inspections, and data management. States are required to match 25 percent of the grant funding allocated with state funding. Funding allocated for EPA-managed class II programs is provided to the EPA regional office to help cover its costs for managing the programs. At least once annually, EPA regional officials are to evaluate state program accomplishments and compare federal funds and state matching funds spent to the state's end of year financial reporting.
- *Aggregates and reports data.* EPA makes data available to the public on the class II program, as well as other UIC programs, nationwide. EPA's guidance says that it will aggregate state-reported data on the UIC program to allow tracking and evaluation of the program. According to EPA officials, EPA developed and maintained a mainframe database to aggregate and summarize state reporting data at a national level until the mid-1990s, and it has not had a similar database until recently when the agency determined that it needed to modernize its data and reporting systems. EPA has three data collection efforts under way: (1) a Web-based system to collect and report basic information on program performance; (2) biannual summary reporting forms submitted by the states or EPA regional offices; and (3) a national well-specific database that, according to agency documents, will ultimately provide data, by well, on a number

of variables, including violations, significant noncompliance, and alleged contamination of underground drinking water. EPA is working with state UIC programs to help them develop the capability to input data directly into the database.

States, and Some EPA Regions, Manage the Day- to-Day Activities of the Class II Program

States, as well as certain EPA regional offices responsible for EPA-managed programs in some states, are largely responsible for the day-to- day management of the class II program. Management includes permitting wells, inspecting wells, enforcing regulations and guidance, and collecting and reporting program data to EPA. More specifically, managing the day-to-day activities of the class II program includes the following:

- *Program management.* States and EPA regional offices are responsible for managing class II program staff, developing program regulations and associated guidance, applying for EPA grants, and tracking expenditures associated with EPA grant funding.
- *Permitting.* States and EPA regional offices review and approve permits for existing and new injection wells and regularly review historical well records.
- *Monitoring and inspections.* States and EPA regional offices review monitoring reports submitted by well operators, conduct field inspections of well sites, test wells to ensure protection of underground sources of drinking water, and ensure that operator reporting, inspections, and testing are done consistently and correctly.
- *Compliance and enforcement.* States and EPA regional offices are expected to identify wells that are not in compliance with regulations and guidance, take enforcement action against well operators in violation of regulations and guidance, and pursue legal action against violators when necessary.
- *Aquifer identification and exemption.* States and EPA regional offices are expected to conduct aquifer surveys and develop inventories of aquifers in the state. They assist with applications to EPA for aquifer exemptions. In addition, they conduct investigations of potential contamination of aquifers.

- *Data management and reporting.* States and EPA regional offices develop a complete inventory of class II wells in the state. They also collect and manage data on wells to satisfy EPA's reporting requirements, including data on well inspections and compliance data, such as well operator violations and any enforcement actions taken.
- *Public information, training, and technical assistance.* States and EPA regional offices conduct public outreach and training to help staff and well operators keep current on technical issues. States and EPA regional offices also provide technical assistance to well operators when necessary.

Resources for State and EPA-Managed Class II Programs Include EPA Grants, State Funds, and Federal and State Personnel

EPA and states provide a mix of resources to support the class II program, including EPA grants, state funding, and EPA and state personnel. State programs are funded in part by EPA grants. States receive EPA grants to help pay for the costs of managing the class II program. Federal officials said that federal grant funds for the UIC program, including EPA grant funds for states' class II programs, have remained unchanged since the 1990s, at around $11 million in grant funding for all states, territories, and tribes. As shown in figure 2, EPA's grant funding for the UIC program, as well as grants for state class II programs, remained flat from fiscal year 2003 through fiscal year 2012.

Between fiscal year 2003 and fiscal year 2012, nationwide, grant funding for state UIC programs has fluctuated between approximately $9.7 million and $10.1 million. During the same period, total UIC grant funding for the states we reviewed remained between $3.0 and $3.2 million, and grant funding for state class II programs was around $4 million.[21] As shown in figure 2, considering inflation and rising costs these totals represent an actual decline in available funds for state UIC programs when converted to real fiscal year 2013 dollars.

State programs also use state funding, including revenues from well permitting fees and other sources, to pay for their programs.[22] For five of the six programs we reviewed that provided budget data, states generally provided most of the funding to support their class II programs (see table 1). Some states have increased their budgets for the class II program over the last few years, while others' budgets have stayed relatively level. Even so, in the states we reviewed, the percentage of the state class II program budgets covered by EPA grants has decreased in the last few years in some states. For example, EPA grant funding comprised approximately 6 percent of California's class II program budget in fiscal year 2008 and approximately 3 percent in fiscal year 2012. Similarly, grant funding comprised approximately 35 percent of Colorado's class II budget in fiscal year 2008 and 30 percent in fiscal year 2012.

Table 1. Select State Class II Program Budgets Fiscal Year 2008 to Fiscal Year 2012

Dollars in thousands					
	State budget				
	EPA grant (% of state budget)				
	2008	**2009**	**2010**	**2011**	**2012**
California	$8,534	$7,572	$12,080	$13,064	$15,460
	$491 (6%)	$494 (7%)	$504 (4%)	$509 (4%)	$489 (3%)
Colorado	$271	$278	$326	$316	$322
	$96 (35%)	$98 (35%)	$97 (30%)	$99 (31%)	$98 (30%)
North Dakota	N/A	N/A	N/A	N/A	N/A
	$97	$99	$103	$105	$100
Ohio	$294	$302	$242	$242	895
	$160 (54%)	$161 (53%)	$161 (66%)	$156 (65%)	$152 (17%)
Oklahoma	N/A	$643	$459	$546	$668
	$286	$291 (45%)	$290 (63%)	$295 (54%)	$287 (43%)
Texas	$2,323	$2,235	$2,174	$2,493	$2,273
	$609 (26%)	$624 (28%)	$622 (29%)	$636 (26%)	$616 (27%)

Sources: GAO analysis of EPA and state data. | GAO-14-555

Notes: State budget data are for the state fiscal year that generally starts in July and ends in June. EPA regions managing programs in Kentucky and Pennsylvania do not track budget data and grant funding specific to the class II programs, while North Dakota could not provide specific data on its program.

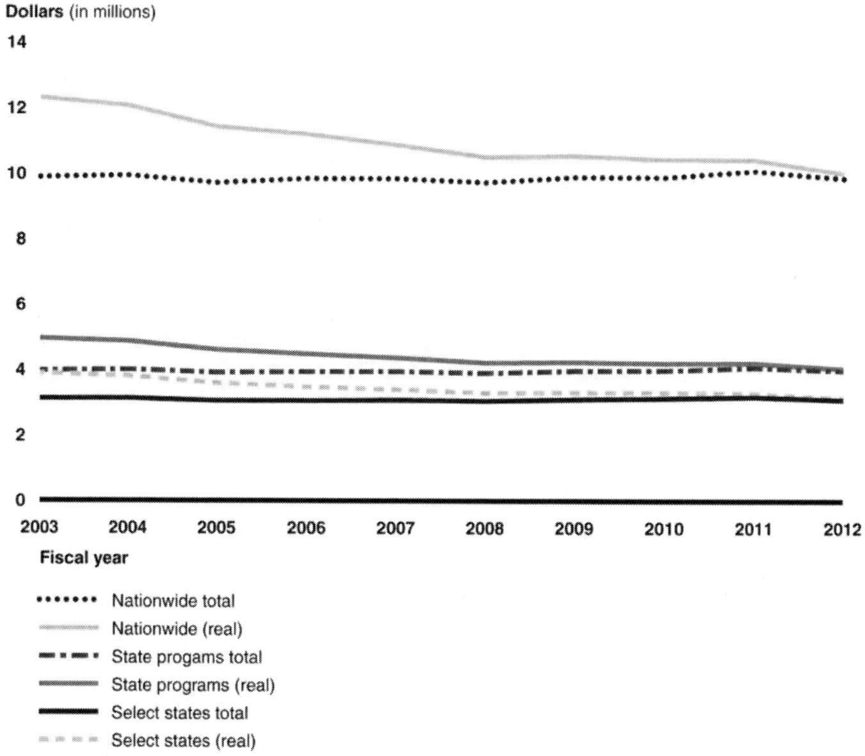

Source: GAO analysis of EPA data. | GAO-14-555.

Figure 2. Federal Funding for State UIC Programs, Including State Class II Programs, Fiscal Year 2003 through Fiscal Year 2012 in Real and Fiscal Year 2013 Dollars.

States also dedicate various staff resources to administer state class II programs. Generally, states utilize management staff to administer the program; technical staff, such as geologists and engineers to review permit applications; and field based staff to conduct inspections and observe well construction. According to state and EPA regional officials, class II programs are managed by a combination of full-time staff and part-time staff that split their responsibilities between the class II programs and other tasks. In the states we reviewed, staffing levels for the class II program have fluctuated over the past several years. For example, Colorado and North Dakota remained flat at approximately 3 and 5 staff members in their programs respectively between 2008 and 2013, whereas Texas and Ohio increased the number of staff in their class II programs from approximately 18 to 24 and 3 to 8 respectively over the same time period. In contrast, the number of class II program staff in Oklahoma decreased from 8 to approximately 6 between 2009 and 2013.[23]

Resources for EPA-managed class II programs, such as Pennsylvania and Kentucky, also include EPA grant funds and EPA staff, although EPA regional offices do not separately track funding and staff resources specific to the class II program. EPA allocates grant funding to all states with UIC programs, regardless of whether they are managed by states or regions, and EPA regions receive this grant funding to help pay for their management. Regions that manage class II programs in certain states said that they have approximately 4 to 6 staff working on the program across all UIC programs, not just the class II program, as of fiscal year 2013. According to EPA officials, staffing resources and funding for the programs has remained relatively flat or decreased since fiscal year 2011. For example, staffing levels for EPA Region 4's class II program in Kentucky have decreased from approximately 8 staff to approximately 6 staff from fiscal year 2011 through fiscal year 2013.

According to EPA and state officials, UIC program responsibilities have increased in the last several years. In particular, EPA, with the input of some state program officials, developed guidance for diesel fuel use in hydraulic fracturing operations and regulations for a new class of injection wells associated with carbon sequestration, known as class VI wells. In addition, between 2005 and 2012, the nationwide inventory of class II wells increased from approximately 144,000 to 172,000.[24]

State and EPA-Managed Programs We Reviewed Have Safeguards to Protect Underground Drinking Water

All of the class II programs we reviewed have safeguards to prevent the contamination of underground sources of drinking water by ensuring that fluids injected into underground formations do not leak into aquifers that are used, or could be used, for drinking water. The six state programs we reviewed incorporate safeguards that EPA deemed effective at preventing underground injections from endangering drinking water sources, while the two EPA-managed programs incorporate the specific safeguards established in EPA regulations and guidance. Officials who manage the eight programs we reviewed reported few known instances of contamination from the injection of fluids into class II wells in the last 5 years; however, EPA's class II program does not require monitoring of groundwater for contamination nor do most of the eight states we reviewed. Moreover, EPA has noted that the absence of known contamination is not necessarily proof that contamination has not occurred. In the last several years, issues have emerged with the potential to affect the protectiveness of class II safeguards, but EPA officials have said that any changes or additions to the safeguards will be addressed on a state-by-state basis.

EPA Safeguards Seek to Prevent Fluids from Leaking into Underground Drinking Water Sources

EPA developed safeguards to protect underground drinking water sources in the 1980s, designing them to prevent fluids that are injected into underground formations from endangering underground drinking water sources. Specifically, in a 1980 document titled *Statement of Basis and Purpose: Underground Injection Control Regulations (Basis and Purpose)*, EPA identified the major pathways that contaminants can take to enter underground sources of drinking water, and discussed EPA's proposed regulations to ensure that movement of injected fluids would not endanger these sources.[25] EPA identified six major pathways of contamination, or ways in which fluids injected into a well could escape the well and enter underground sources of drinking water. Figure 3 shows four of the six different pathways.

According to EPA, the pathways include

1. fluid movement through a hole or other fault in the well's casing, or steel pipe that is placed into the wellbore—that is, the hole in which the well is placed. Casing can prevent a well hole, or bore, from collapsing and, in specific cases, also serves as a means for injecting fluids into the underground formation in which they will be stored.
2. fluid movement through the space between the casing and the wellbore. Such movement can occur when friction and resistance are created in the formation into which fluid is being injected, and the fluid takes the path of least resistance back through the casing and the wellbore.

3. fluid movement from an injection zone, or the underground formation into which it is injected, through the confining formations around it. When they are injected into a formation, fluids under pressure will normally travel laterally through the formation, away from the well. Typically, the formation should be separated from overlying formations that contain drinking water by a confining layer, or a layer of relatively impermeable rock; however, if there are permeable or fractured areas in the confining layer, fluids can move from the injection formation into a formation that serves as a source of drinking water.
4. fluid movement through abandoned or completed wells that are not properly plugged to prevent movement of fluid. This occurs when fluids injected into a formation move laterally through the formation and encounter a well that has been abandoned and not properly plugged or a well that is complete and operating, but has not been properly completed (i.e., weaknesses are present). Fluids injected under pressure will take the path of least resistance and flow up the wells and into underground formations containing sources of drinking water or even onto the land surface.
5. fluid movement from one part of a formation to another that is not meant to be used for wastewater storage. In some cases, fluids may be injected into an aquifer that in another area is designated for use as a drinking water source. In these instances, the injection formation does not have any confining layers or other geologic formations to separate it from the drinking water source. According to EPA officials, injection is typically done when the flow of an aquifer is away from the protected part of the aquifer, or when the injection pressure is low enough to prevent movement to the protected part of the aquifer.
6. fluid injection directly into a drinking water source. This occurs when fluid is injected directly into an underground source of drinking water. The Safe Drinking Water Act and EPA regulations prohibit injection of contaminants directly into an underground drinking water source, if the presence of that contaminant may cause a violation of any primary drinking water regulation or otherwise effect human health.

To prevent fluids from moving along these pathways and potentially contaminating underground sources of water, EPA designed several safeguards described in the *Basis and Purpose report*. These safeguards, shown in figure 4 and described below, are flexible and implemented by the state program directors. EPA has not reviewed class II program safeguards since the 1990s, but officials told us that, generally, the safeguards established at the UIC program's inception remain sufficient to ensure the protection of underground sources of drinking water. In the eight states we reviewed, two states' programs (Kentucky and Pennsylvania) are managed by their respective EPA regions and have adopted the EPA safeguards. The six remaining states we reviewed have safeguards in their programs that EPA has deemed protective of underground sources of drinking water. These states were not required to use the safeguards in EPA regulations. Rather, EPA reviewed the state programs' safeguards to determine whether they were effective in protecting underground sources of drinking water. In the following discussion, for each safeguard, we discuss EPA regulations, which apply only to Kentucky and Pennsylvania, and then we discuss how the other six state programs address each safeguard.[26]

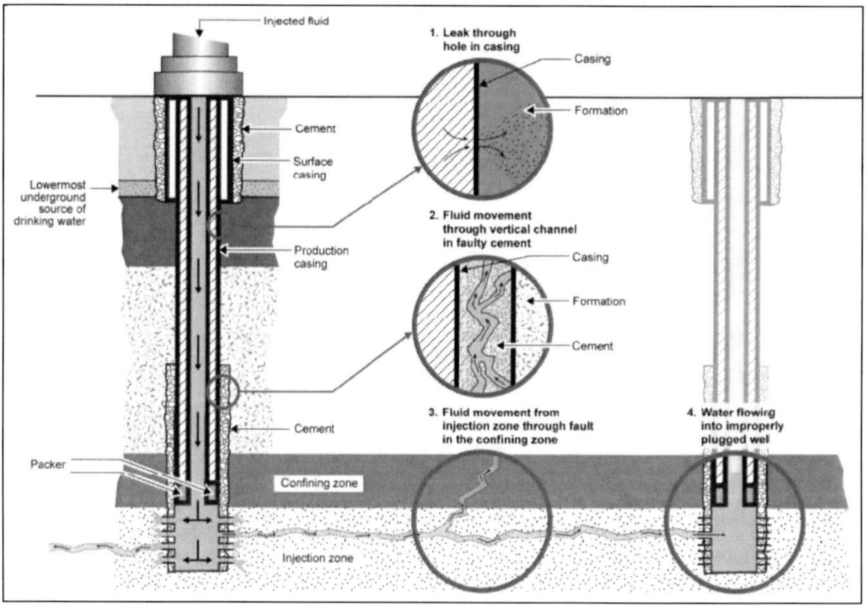

Source: GAO analysis of EPA information. | GAO-14-555.

Note: Other pathways that are not included in the graphic include fluid movement from one part of a formation to another that contains an underground source of drinking water and fluid injection into a drinking water source.

Figure 3. Pathways of Contamination of Underground Sources of Drinking Water by Class II Wells.

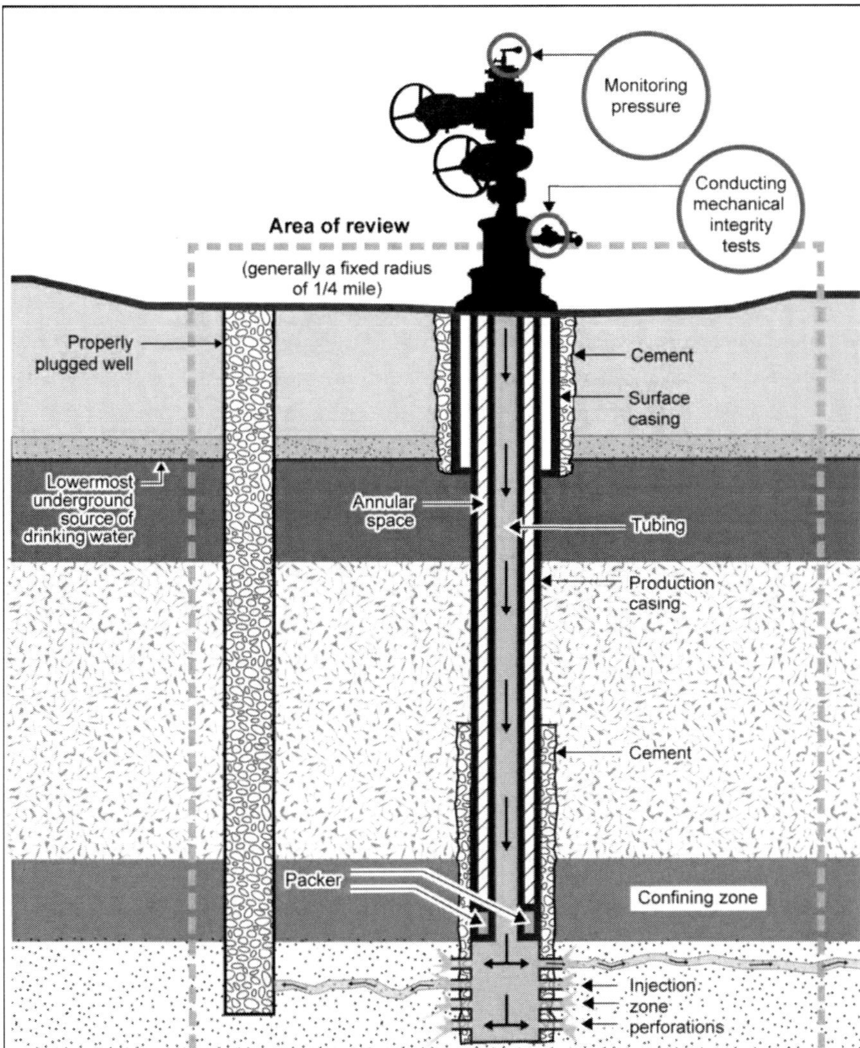

Source: GAO analysis of EPA and Ohio's information.| GAO-14-555.

Figure 4. Class II Injection Well with Safeguards.

Area of review: To prevent fluids from entering an underground drinking water source by, for example, moving through wells that have been improperly abandoned or completed, the state UIC program director determines an area of review necessary to obtain a permit for new injection wells. This is an area around an injection well where pressures in the injection zone may cause the migration of fluids into an underground source of drinking water. Before fluids can be injected into a new well, the state director must consider information on other active or abandoned wells in the area, and the corrective action status of any defective wells. The size of an area of review can be a fixed radial distance of one-quarter mile or greater, or it can be calculated by a formula that considers the injection rate, the movement of fluids through the injection zone, and the size of the injection zone among other factors. The area of review that results from the calculation is known as the zone of endangering influence, and can be greater or smaller than the fixed quarter mile radius. As shown in table 6 in appendix IV, all eight programs we reviewed generally require that at least an area of one-quarter mile around a well be reviewed to ensure that wells are properly abandoned or completed. In addition, class II program officials in three of the states we reviewed, noted they have the discretion to require that the zone of endangering influence be calculated to determine the area of review. Some state programs also have exceptions to the fixed radius requirements for the area of review. For example, in Ohio, if operators plan to inject into a well at a rate greater than 200 barrels per day, the fixed radius of the area of review increases to a half mile. In Oklahoma, if a well can accept fluids via gravity and does not require additional pressure to inject fluids, no area of review is required.

Geologic characteristics of injection zone and confining layers: To ensure that fluids do not travel through weak areas of a confining layer, EPA regulations require that appropriate geological data on the injection zone and confining zone be considered prior to issuance of a permit to inject fluids. Programs can use data that already exist, and a permit cannot be issued until a determination has been made that the formations are sound and can contain injected fluid. In the states we reviewed, permit applications require prospective well operators to provide basic information on geologic formations surrounding the injection zone including factors such as the permeability and porosity of the surrounding formation and the depth of the lowest freshwater bearing formation. In all of the states we reviewed, permit applications require some geologic information on the formation where oil and gas fluids were being injected. In addition, all but one of the states we reviewed collect information on the depth of underground sources of drinking water or freshwater in the area surrounding the injection zone.[27], [28] Table 7 in appendix IV shows the information to be submitted for geologic review prior to permitting for the eight states we reviewed.

Casing, cementing, tubing, and packer: To prevent fluid from moving through a faulty injection casing, EPA set specific requirements for well construction. According to EPA's Basis and Purpose report, some wells only need surface casing, or casing that extends the length of the bore through the formation in which shallow drinking water exists, while other wells may need multiple sets, or "strings," of casing depending on the depth of the well and the surrounding geologic formation. For example, intermediate casing can be necessary to protect other underground resources such as coal beds or gas storage zones. The deepest layer of casing, known as the long-string or production casing, isolates injection of fluids into the designated formations. EPA also requires that each layer of casing be surrounded by cement and suggests that specific equipment called tubing and packer be used. Tubing is typically steel or plastic pipe inserted inside the production casing, which isolates the casing from the fluid injected into the well. Tubing set on a packer allows well operators to directly inject fluids into formations and prevents corrosion by not allowing injected fluids to contact the casing wall. A packer is a mechanical device that sits below the tubing and locks into the casing wall, sealing the space between the tubing and casing, called the annulus, from the injection zone. In Kentucky and Pennsylvania, EPA manages the class II program using EPA's requirements. Each of the six state programs we reviewed have requirements for casing and cementing, with varying levels of specificity on depths of casing and cementing depending on the location of underground sources of drinking water or well type, among other factors. For example, North Dakota regulations state that all casing and cementing requirements should be dependent on the geological factors in the area surrounding the well. Conversely, California and Ohio have specific requirements regarding depth of cementing for surface, intermediate, and production casings. Table 8 in appendix IV shows each state's requirements for casing and cementing.

Mechanical integrity testing: To prevent fluids from leaking out or up through the wellbore, along the outside of the cement surrounding the casing, EPA's Basis and Purpose report states that a well needs to demonstrate mechanical integrity, or the absence of leaks. To ensure this, EPA's regulations require that the integrity of well casing be verified through, among other means, mechanical integrity testing to detect leaks. Mechanical integrity testing involves, for example, increasing the pressure in the tubing and ensuring that the tubing and packer holds that pressure for a period of time. Verification of mechanical integrity first occurs prior to the well being authorized as ready for injection, with subsequent verification occurring at least once every 5 years during operation for most wells. All of the states we reviewed, except for Ohio, generally require that new injection wells be tested prior to injection and that they be tested at least once every 5 years of operation. Ohio requires mechanical integrity of wells before injection of fluids commences and then requires monthly monitoring for leaks in saltwater disposal and enhanced recovery wells; if monthly testing is not feasible, then these wells must be tested for mechanical integrity every 5 years. California and Oklahoma also require more frequent testing of mechanical integrity for specific types of wells. For example, California requires annual mechanical integrity testing for every water disposal well, and Oklahoma requires annual testing for all commercial disposal wells. Table 9 in appendix IV shows the requirements for mechanical integrity in each of the eight states we reviewed.

Injection pressure: Another measure to ensure that fluids do not travel through the confining layers into a source of drinking water is the control of the pressure at which fluids are injected, or injection pressure. EPA regulations require that well injection pressure should be controlled to avoid initiating new fractures or propagating existing fractures in the confining zone adjacent to underground sources of drinking water. In the eight states we reviewed, permit applications require information on the rate of injection and injection pressure to determine the maximum allowable injection pressure. Table 10 in appendix IV shows the injection pressure requirements for each of the states we reviewed.

Plugging and abandonment: To prevent fluid from moving through improperly abandoned wells, EPA regulations require that, after operation of a well stops, the wellbore is adequately plugged with cement. Each of the eight states we reviewed had requirements for injection wells to have a plugging and abandonment plan and sufficient financial resources, usually in the form of a bond, to complete the plan if necessary. Table 11 in appendix IV shows the plugging and abandonment requirements for each of the eight states we reviewed. All of the states we reviewed except North Dakota require that state or EPA officials witness well plugging or that operators report to the state once the well has been plugged.[29] State requirements for financial assurance varied between the states we reviewed depending on the depth of the well, well status, and the number of wells covered. For example, California, Colorado, and North Dakota increase the financial resources, or bonding amount, required based on the depth of the well. In addition, California and Colorado also require financial resources for the plugging and abandonment of inactive wells. Some states will also allow operators to obtain an individual bond for multiple wells.[30] In Kentucky and Pennsylvania, where the class II program is managed by EPA regions, the regional administrator determines the financial resources necessary for an operator to plug and abandon a well, and the operating company must demonstrate that it has those resources available.

Further, in some of the eight states we reviewed there are old wells that were abandoned and not plugged properly and, in some cases, without an operator to hold financially responsible. In addition, according to EPA regional and state officials, some states are unaware of the location of some of these wells. For example, EPA officials said that the oil and gas industry in Kentucky has been present in the state for more than 100 years and that improperly plugged wells predating the class II program pose a key vulnerability to underground sources of drinking water. According to EPA officials, many of these wells were plugged with tree stumps or other readily available material and have insufficient casing or cementing. California, Kentucky, Pennsylvania, and Texas have a backlog of abandoned wells that may need to be plugged to protect underground drinking water sources. In some cases, if the abandoned wells fall within one-quarter mile of the proposed new well, the states can require that operators plug the abandoned wells through the area of review requirements for new injection wells. They can also use other resources to pay for plugging these wells. Of the eight states we reviewed, seven reported having an abandoned well fund from which they can pay to plug and properly abandon a certain number of wells per year.[31]

The eight states we reviewed also have specific requirements for wells where injection will not take place for several months or years, defined by EPA as temporarily abandoned wells (see table 12 in app. IV). Programs in the eight states we reviewed have different definitions and requirements for wells that are not in use before the well is plugged, and do not require additional testing for wells that are inactive. EPA guidance states that a uniform definition of temporary abandonment is not needed as long as requirements are in place to ensure that wells are not endangering underground sources of drinking water.

Monitoring and reporting: To provide an early warning of potential problems, EPA regulations require monitoring of fluids to be injected and well operation. For class II wells, EPA requires, among other things, that operators "monitor the nature of the injected fluids with sufficient frequency to yield data representative of their characteristics;"[32] according to EPA's Basis and Purpose report, such information can help federal and state regulators understand reasons for well failures and take appropriate corrective actions. In addition, class II wells should be monitored on a daily to monthly basis; according to EPA, continuous monitoring is not required for class II wells because the fluids injected into these wells are usually less corrosive and less hazardous than fluids injected into other UIC well classes. In each of the states we reviewed, programs require class II well operators to report on injection pressure and injection volume to ensure the well is being properly monitored. The programs we reviewed vary in the frequency and type of reporting required based on the type of well. For example, California, Colorado, and North Dakota require monthly reporting on injection pressure, injection volume, and the type of fluid being injected whereas Kentucky, Ohio, Pennsylvania, and Texas require annual reporting. Table 13 in appendix IV shows the monitoring and reporting requirements in the eight states we reviewed.

According to EPA guidance, inspections of well sites are important to ensure that operators have safeguards in place to effectively protect underground sources of drinking water. Programs in each of the states we reviewed conduct on-site inspections to review the operation and condition of the wells. If these inspections identify violations of the applicable federal or state regulations safeguarding underground sources of drinking water, the state can proceed with enforcement activities, such as fines and penalties, as appropriate. Table 2 shows the inspection activity in state fiscal year 2012 for the eight states we reviewed compared to the class II inventory of wells for 2012. The percentage of wells inspected compared to the state's total well inventory ranged from 20 percent in Oklahoma to 100 percent in Ohio and North Dakota.

Table 2. Inspections of Class II Underground Injection Control (UIC) Wells in Selected States

State	Class II well inventory	Actual inspections in fiscal year 2012	Percentage of class II wells inspected in fiscal year 2012 as reported by states[a]
California	49,783	31,869	64 %
Colorado	901	769	87
Kentucky	3,221	1,127	34.3
North Dakota	1,290	6,778	100
Ohio	2,439	1,686	100[b]
Oklahoma	11,134	11,680	20
Pennsylvania	1,865	330	33
Texas	52,977	22,412	41.4

Sources: Inventory data reported by EPA; inspection data reported by state officials.

[a] States may inspect a well more than one time in a fiscal year, and the percentage of wells states reported inspecting may not reflect the number of actual inspections divided by total well inventory. Conversely, states may inspect more than one well during the same inspection. At GAO's request, select states calculated the percentage of class II wells inspected for fiscal year 2012. In calculating their percentages, it is possible that some states may have counted multiple inspections to a single well multiple times in this calculation, while other states may have removed them as duplicates.

[b] According to EPA Region 5, Ohio's inventory of class II wells includes approximately 2,000 temporarily abandoned annular disposal wells which are not annually inspected. The actual inspections column only reflects inspections conducted on the approximately 440 active conventional and annular disposal wells in Ohio.

Programs in States We Reviewed Reported Few Instances of Alleged Contamination, Although Class II Program Safeguards Do Not Include Groundwater Monitoring

The programs in the eight states we reviewed reported few instances of alleged contamination caused by potential leaks from underground injections into underground drinking water sources. State and EPA officials reported this information from two sources: (1) data on well violations that could be significant enough to contaminate underground sources and (2) data on citizen complaints of water well contamination and resulting state investigations.

States and EPA-managed programs track data on well violations that are significant enough to pose a risk of contaminating underground sources of drinking water. State and EPA regional officials identify violations through their well inspections and, using EPA definitions, identify the violations that may have been significant enough to contaminate underground sources of drinking water. State and regional officials do not have to confirm that contamination has occurred, only that a violation was significant enough that fluids may have contaminated an underground source of drinking water. Officials for the eight programs we reviewed said that they reported few such significant violations, and that few of these had actually contaminated drinking water sources, as shown in table 3. For example, California reported 9 instances of alleged contamination in 2009, and 12 instances of alleged contamination in 2010; California officials told us that the instances of alleged contamination resulted from an individual operator that was injecting fluids illegally into multiple wells.

Table 3. Instances of Alleged Contamination in Selected States from Class II Underground Injection Control (UIC) Wells

State	2008	2009	2010	2011	2012
California	0	9	12	0	3
Colorado	0	0	0	0	0
Kentucky	0	0	0	0	0
North Dakota	0	0	0	0	0
Ohio	0	0	0	0	0
Oklahoma	1	2	4	1	2
Pennsylvania	0	0	0	0	0
Texas	0	1	0	0	0

Source: GAO analysis of EPA 7520–2A data.

Notes: According to California officials, all of the instances of alleged contamination California reported in 2009, and 9 of the 12 instances of alleged contamination reported in 2010, resulted from one operator injecting illegally into multiple wells. The instances of alleged contamination may vary because of several factors, including the number of inspections that each state conducts, how states prioritize inspections, and how states define alleged contamination.

Officials from the eight states we reviewed also based their statements about few or no instances of contamination from injection wells on their efforts to track and investigate citizen complaints alleging water well contamination. Each of the eight states we reviewed has a process for receiving, tracking, and investigating citizen complaints about their water wells being contaminated. The process the states use is similar, taking citizen complaints and investigating the complaints. The investigation may involve, for example, checking the drinking water well for contamination and completing an assessment of the possible sources of contamination.

However, unless it is part of an investigation of a citizen or other complaint, EPA and states generally do not directly monitor groundwater to detect contamination from injection wells as part of their class II programs. When it first developed the UIC program and its regulations, EPA considered, but did not include, monitoring of groundwater for contamination as a means of evaluating the effectiveness of the program and its safeguards.[33] Furthermore, the Safe Drinking Water Act does not specifically require groundwater quality monitoring for class II wells. Moreover, EPA guidance notes that, while evidence of the presence or absence of groundwater contamination is important, it cannot serve as the only measure of program effectiveness, and the absence of evidence of contamination does not necessarily prove that contamination has not occurred.

Nonetheless, some of the programs in the eight states we reviewed do monitor baseline water quality, although EPA's regulations do not require it. North Dakota, Oklahoma, and Colorado each require sampling of groundwater at various stages of the oil and gas production process. For example, North Dakota requires applicants for a class II injection well permit to provide a quantitative analysis from a state-certified laboratory of freshwater from the two freshwater wells nearest to the proposed injection well. Similarly, Oklahoma requires permit applicants to provide an analysis of freshwater from two or more freshwater wells within a one-mile radius of the proposed injection well. Colorado officials said the state requires oil and gas drilling permit applicants to analyze the water quality of groundwater samples from four nearby water wells, depending on well location and depth.

New Program Risks Have Emerged that EPA Says Will be Handled on a State-by-State Basis, but State Programs Could Benefit from Additional National Guidance

New risks have emerged in the management of class II disposal wells that could affect the class II program: induced seismicity, overpressurization of formations, and use of diesel fuel in hydraulic fracturing operations. EPA has tasked the UIC Technical Workgroup with providing technical information to inform states' decisions about induced seismicity, but plans to address overpressurization of formations and diesel use on a state-by-state basis without requesting assistance from the workgroup. Without similar reviews of other emerging risks, class II programs may not have the information necessary to fully protect underground drinking water.

Induced seismicity: Recent seismic activity associated with injection wells in Arkansas, Ohio, Oklahoma, Texas, and West Virginia has raised awareness of the potential for earthquakes resulting from the underground injection of fluids. In addition, in 2012, the National Academy of Sciences concluded that underground injection does pose some risks for induced seismicity.[34] To the extent that induced seismicity creates conditions that endanger underground sources of drinking water, according to EPA, it may have negative environmental effects.[35] Programs in three of the eight states we reviewed—Ohio, Texas, and Oklahoma—have adopted steps to address the potential for seismic activity in injection areas. Ohio finalized new regulations that went into effect on October 1, 2012 that allow the state's chief of the Division of Oil and Gas to require a number of different tests or evaluations to address potential induced seismic risks for companies seeking permits for brine injection wells in Ohio.[36] According to Region 6 officials, Texas hired an in-house seismic expert to assess potential risks and the state continues to monitor developments and research related to induced seismicity. Oklahoma has partnered with Region 6 to conduct three dimensional mapping of seismic events for analysis.

EPA officials said the agency has not amended its regulations to add specific requirements related to seismic activity and injection wells, but rather, tasked its UIC Technical Workgroup with conducting a study of the problem and potential actions to be taken by EPA and the states. The workgroup issued a draft white paper that identified the three key components behind injection-induced seismic events and identified possible steps to be taken by state programs to manage or minimize induced seismicity, including (1) determining whether a site needs further assessment to ensure protection of underground sources of drinking water; (2) taking steps to assess the reservoir and to modify well operations (injection pressure, intervals, or other measures); and/or (3) require additional seismometers or increase monitoring of injection pressures, formation pressures, and/or the characteristics of the fluids being injected. EPA officials said that the white paper is still in draft, but will help the state program directors to decide what, if any, regulations or safeguards need to be adopted to deal with the issue of induced seismicity.

Overpressurization of formations: Overpressurization, according to Region 6 officials, occurs when fluids injected into a formation increase the pressure in it to a point where the fluids flow back up a well and onto the surface. Region 6 officials said that two such incidents occurred in Oklahoma, and they noted that overpressurization is occurring in locations where formations have been developed and receiving wastewater for long periods of time. The key threat to underground drinking water sources in such situations is from the leaked fluids containing contaminants from produced water or fracturing fluids flowing to the surface and migrating back into formations containing underground sources of drinking water. Region 6 officials indicated that fluids could leak from the well into groundwater formations through other pathways and that it is difficult to determine if this has happened. In 2003, Region 6 held a conference, where officials expressed concern that a large number of injection zones were becoming overpressurized because they were seeing an increase in requests for permit modifications to increase injection pressure.[37] According to Region 6 officials, no additional actions were taken to address the region's concerns regarding overpressurization. EPA officials said that instances of overpressurization occur infrequently and that they plan to address overpressurization of formations on a state-by-state basis, not through the UIC Technical Workgroup. The issue could affect other states, however, as increased volumes of fluids are injected into formations; these states and EPA regions could benefit from the information that other states and regions have learned.

Diesel fuels: The use of diesel fuels as chemical additives in oil and gas production—specifically in hydraulic fracturing of shale or similar formations to stimulate production—has raised concerns over the risks to underground sources of drinking water. Diesel fuels are sometimes used as components of hydraulic fracturing fluid and contain toxic compounds such as benzene and other aromatic compounds that can pose environmental and human health risks.[38] In February 2014, EPA issued program guidance accompanied by a memorandum to EPA regions saying that under EPA-managed class II programs injections of diesel fuels for hydraulic fracturing are subject to permitting requirements.[39] The guidance is intended to provide technical recommendations for protecting underground sources of drinking water from potential endangerment posed by hydraulic fracturing with diesel fuels, and it includes nonbinding recommendations for EPA regions to consider in applying class II regulations to these injections. For example, whereas EPA's class II regulations require that a well owner or operator provide state or EPA-managed programs with an appropriate analysis of the chemical and physical characteristics of the fluids to be injected, the guidance suggests focusing on prepermit water quality monitoring in the area of review.[40]

According to EPA and state program officials, the UIC program guidance for diesel use in hydraulic fracturing will be implemented on a state-by- state basis and does not need review from the UIC Technical Workgroup. However, while it is the responsibility of the operator to obtain a permit for any injection covered by UIC program laws or regulations, the information officials need to ensure that diesel permits are issued when necessary may not be available, depending on state requirements and practices. Specifically, officials that manage seven of the eight state programs we reviewed said that diesel fuel, as defined by EPA's guidance, are not now being used in oil and gas production in their respective states, and none of the states we reviewed have issued permits for use of diesel fuels in hydraulic fracturing, according to officials. California did not know whether diesel had been used or not. To discover whether companies were using diesel fuel, some of the states said that they had reviewed available information on chemical disclosures and discussed operations with oil and gas companies to determine whether companies are using diesel fuels in hydraulic fracturing operations. While several states have begun to require well owners or operators to use a national reporting system called FracFocus to disclose chemicals used in hydraulic fracturing fluid that could help states identify hydraulic fracturing operations using diesel fuels, not all states have done this, which means that all operators may not be providing information, and the information available is not complete.[41] Furthermore, operators consider some information, such as hydraulic fracturing fluid chemical composition, to be classified business information, which is not subject to public disclosure. Without an assessment of the complete chemical information needed for permitting, such as an assessment by the UIC Technical Workgroup, EPA and the states may not have the chemical disclosure information they need to ensure permits are issued for wells that use diesel fuel in hydraulic fracturing.

EPA Is Not Consistently Performing Oversight and Enforcement Activities

EPA is not consistently performing two key activities associated with its oversight and enforcement responsibilities for class II programs. First, EPA does not consistently conduct annual on-site reviews of state programs, which EPA guidance identifies as a key activity needed to conduct effective oversight and to ensure that state and EPA-managed class II programs are protecting underground sources of drinking water. Second, EPA is not consistently incorporating changes to state class II program requirements into federal regulations, as required by its regulations, to enable enforcement of state program requirements if necessary.[42]

EPA Is Not Consistently Conducting Annual On-site Evaluations of State or EPA-Managed Programs

As part of effective oversight of state programs and EPA-managed programs, EPA's guidance recommends that regional staff and headquarters staff, conduct several ongoing oversight activities. These include (1) reviewing annual reports from states, (2) reviewing financial reporting on grant funding from state programs, (3) reviewing state reports on injection wells that do not comply with federal or state regulations, and (4) conducting annual on-site program evaluations.

For the regions we reviewed, EPA regional officials regularly conducted three of the four oversight activities identified in EPA guidance. For example, in the eight regions we reviewed, EPA officials regularly reviewed states' annual reports and forms identifying noncompliance within each state to identify any areas of concern and followed up with state officials to discuss and resolve them. In addition, EPA regional officials reviewed grant financial reporting for class II programs as part of their annual grant program review.

However, EPA regions we reviewed did not consistently conduct annual on-site program evaluations as directed in the EPA guidance, nor did EPA headquarters conduct such evaluations of EPA-managed programs. According to EPA guidance, EPA regions should perform at least one on- site evaluation of each state program each year to, among other things, assess whether the state is managing the program consistent with state regulations, setting program objectives consistent with national and regional program priorities, and implementing recommendations from previous evaluations. Following the on-site evaluation, the regional office should draft a written report on the state's performance and submit the report to EPA headquarters and the relevant state program office. EPA headquarters is responsible for conducting similar program evaluations of EPA-managed programs and producing the associated written reports.

EPA officials recognize the benefits of on-site evaluations of state programs, but said they have limited resources to conduct them. Regional officials said that on-site program evaluations are valuable for coordinating between federal and state officials to improve program management. For example, a comprehensive evaluation of the California class II program contracted by Region 9 in 2011 resulted in a number of recommendations, and as a result, California is planning to update their regulations and hired 43 additional staff in the division, including additional staff responsible for managing their class II program to bolster regulatory activities. In addition, officials in Region 7 and Region 10 told us that their evaluations have identified deficiencies in the financial requirements of well operators planning to drill new class II injection wells that are necessary to ensure that those wells can be adequately plugged in an emergency. However, according to EPA

program regulations.[45] As of May 2014, EPA Region 6 has not officially approved these changes and incorporated them into federal regulations, which Oklahoma submitted in draft to EPA in 2011. According to Region 6 officials, regional personnel have not reviewed or approved Oklahoma's program changes because other regional responsibilities, such as coordinating the induced seismicity white paper, are a higher priority given the agency's limited resources. Furthermore, in two states, EPA never incorporated any state program requirements into federal regulations. Specifically, while EPA approved the class II programs in West Virginia and Arkansas in the 1980s, the agency never incorporated any of the states' program requirements into federal regulations and, as a result, does not have enforcement authority for class II program regulations in either state.[46]

Without incorporating state program requirements, or changes to state program requirements, into federal regulations, EPA may not be able to take enforcement action, if a state does not take such action or requests EPA's assistance to take action, against well operators that are violating state regulations. Under EPA regulations, the agency cannot enforce regulations that it has not approved by rule.[47] For example, according to an EPA official, Illinois requested that EPA Region 5 enforce the state's class II UIC requirements against an Illinois well operator for violating the regulations after the state was unsuccessful at getting compliance from the well operator through its own enforcement actions. The operator did not conduct required mechanical integrity tests on six injection wells and did not submit annual status reports for the wells. Without the operator's compliance with testing and reporting requirements, Illinois program officials were unable to determine whether the wells were at risk of contaminating underground sources of drinking water. EPA originally issued an administrative order assessing a fine of $105,000. According to EPA officials, the operator challenged EPA's enforcement action several times over a period of nearly 10 years, and eventually appealed the case to the U.S. District Court. While on appeal, EPA discovered that the latest Illinois regulations had not been incorporated into the federal regulations. EPA moved to remand the case, and in 2012, the court granted EPA's request to remand the underlying decision of EPA's Environmental Appeals Board,[48] and EPA later settled the case with the estate of the operator for $20,000. EPA officials said that they do not often have to step in to enforce a state class II program regulation, but as the oil and as industry continues to develop its resources and use innovative technologies, state programs may change their regulations, and EPA may have increasing numbers of state program changes to review, approve, incorporate into federal regulations, and enforce.

Because EPA has not been incorporating changes to state program requirements into federal regulations, the agency has a backlog of state program requirements that it cannot enforce if necessary. When faced with a similar backlog in the early 1990s, EPA conducted a review to identify all state program changes since the UIC program's inception in the early 1980s, and then conducted one large rulemaking to incorporate all of the identified changes into federal regulations. EPA spent 3 years researching and comparing state regulations to those already incorporated into federal regulations, and it identified changes to 37 state programs that needed to be incorporated into federal regulations through its rulemaking. However, EPA has not undertaken a similar effort to identify and incorporate changes to state program requirements into federal regulations since 1991. In 2010, EPA UIC officials assessed the resources that would be needed to conduct a similar effort, and they estimated the time and resources necessary to complete those steps. At that time, according to EPA documents, EPA could not verify that any state programs were up-to-date in federal regulations, and estimated that it would require 2 to 3 years, $150,000 in outside contract support, and dedicated EPA staff to identify and incorporate all state program changes made since 1991 into federal regulations. Until it conducts a rulemaking to incorporate the backlog of state program requirements and changes to state program requirements that have been approved, EPA will not be able to enforce some state program requirements, hindering the agency's enforcement of the program nationally.

EPA officials told us that incorporating changes into federal regulations, particularly through the rulemaking process, was burdensome and time-consuming. Several EPA officials told us that reviewing, approving, and then incorporating changes into federal regulations through rulemaking is lengthy and resource intensive. As a result, the agency has not conducted rulemakings to keep pace with the changes that are occurring in state programs. The requirement that EPA incorporate changes to state regulations into federal regulations is established through the agency's regulations, however, and is not required by the Safe Drinking Water Act. According to EPA officials, other EPA programs have a less burdensome process for reviewing and approving changes to state programs that does not require a rulemaking for approval, and does not require a separate rulemaking process for EPA to incorporate changes into federal regulations. For example, under EPA's Public Water System Supervision program, the EPA Administrator can review and approve changes to state programs and maintain enforcement authority, and the agency does not require that those changes be incorporated into federal regulations.[49] The agency has discretion under the Safe Drinking Water Act to change its UIC regulations to revise or eliminate the requirement for incorporating state program changes into federal regulations, but, according to officials, has not evaluated alternatives to its current approval process. For example, EPA has not evaluated whether it could remove the requirement for a separate rulemaking to incorporate state program changes, or somehow modify its approval process to ensure that state program changes are incorporated into federal regulations at the same time changes are reviewed and approved.[50] Until EPA evaluates whether this requirement can be revised or eliminated to make the review, approval, and incorporation of state program changes more efficient, the process for incorporating future state program changes will remain burdensome and time-consuming.

EPA Collects a Large Amount of Data on Class II Wells, but the Data Are Not Sufficiently Complete or Comparable to Report on the Nationwide Program

EPA collects large amounts of data on class II wells in each class II program, but the data are not sufficiently complete or comparable for reporting to Congress, the public, or other groups interested in the nationwide program. According to EPA's guidance, the agency will establish a tracking and evaluation system for the program, and provide the Congress and other groups with information to assess the program. To this end, the agency collects data on class II programs across the country using required activity reports from state programs, and, to a lesser extent, a Web-based performance management database. Our review of the data shows that it is not sufficiently complete or comparable to report on the program at a national level.

EPA Collects a Large Quantity of Data on Class II Programs

To satisfy EPA's goal of aggregating and reporting state data for the purpose of responding to inquiries from Congress and other interested groups,

EPA collects a large amount of summary data on class II programs. It does so using the following two methods:

- *Activity reporting*. EPA's primary method for collecting summary data on state UIC and class II programs is its series of activity reporting forms known as the 7520 forms. These forms are collected twice a year from state programs and EPA-managed programs. The states are to provide the forms to their respective EPA regional offices, and the regional offices are supposed to submit the forms to EPA headquarters for review. Five forms comprise the core of the 7520 series; each form collects information at a summary level on injection wells under a state's program, reported by type of UIC well, including class II wells. Data collected include summary information on the total number of permits issued during a year, total violations cited by inspectors, enforcement actions taken on wells that do not comply with regulations, wells that are repaired and returned to working order—a process called returning to compliance–within specified times, and total number of inspections and testing of the mechanical integrity of wells.
- *Performance reporting*. In addition to its 7520 forms, EPA collects summary data on two basic performance measures through a Web-based program known as the Inventory Measures Reporting System, to satisfy requirements of the Government Performance and Results Act.[51] The Inventory Management Reporting System was created in 2003 and collects data on the inventory of wells—that is, the number of each type of injection well per state, including class II wells. This inventory data is one factor used by EPA to calculate states' annual grant allocations. In 2006, the function of this Web-based system was expanded to also collect information on the number of wells that failed mechanical integrity tests and returned to compliance within 180 days (the number of days that EPA deems appropriate). Both pieces of data are generally entered into the Web-based system by EPA regional officials; EPA reports the data by state in its annual performance report.[52]

Because the paper forms it uses are burdensome and time-consuming to pull together, according to EPA officials, the agency is focusing its efforts on the creation of a national UIC database to be able to report on the national program. This database is designed to collect well-specific data from state agencies—as opposed to summary data—and from regional offices in cases where EPA manages state programs. The database will also receive data electronically directly from the states and regions, rather than through paper documents sent from the states to EPA regions, and then to the agency's headquarters. The database has been under construction since 2007, and EPA officials report that it is now functional; however, this database is not fully populated, as of January 2014, with only eight class II programs uploading data into the database. EPA officials told us that the database will not be complete and widely populated enough to report national data for at least 2 to 3 years. Until the agency has a fully populated database, EPA will not have a ready way to aggregate and report data on the UIC program, and class II wells in particular. Without this reporting, EPA may not fulfill its goal of providing information on the national program. Given public concerns over oil and gas development generally, demand for such data may increase; however, in light of EPA's budget limitations, it could be longer before the UIC database is complete and the data are available.

EPA's Data Are Not Sufficiently Complete or Comparable for Reporting on the Class II Program at a National Level

The data that EPA collects from its 7520 forms represents the most detailed and extensive set of data the agency compiles on the program; however, it is not sufficiently complete or comparable to allow EPA to aggregate state information and report on the national class II program. The forms are not complete for two reasons. First, the forms are generally received in paper format making the data difficult and time-consuming to summarize and report.[53] As a result, EPA headquarters does not have all the 7520 forms to summarize and report. We requested, but were unable to obtain, a full 5 years of 7520 forms for all states. EPA officials told us that EPA headquarters provided all of the forms available, but that, in some cases, the forms had not been submitted by the states. The agency has attempted to compile the data as part of an effort to put the data in a simple electronic spreadsheet, but it acknowledges that it is missing state data and would have to get it from the regional offices or individual states. Second, in our review of the 7520 forms provided to us by EPA, we found examples of incomplete forms, including forms without state names and forms with blank fields, with no indication of whether the blanks represented missing data or a quantity of zero. According to officials at the regional level, there are no data entry protocols to indicate whether these blank fields represent zeros or missing data. While EPA regional officials reported that they were familiar enough with operations in the states that they knew which blanks were supposed to be zeros and which were supposed to be blank fields, none reported editing the forms before transmitting them to headquarters. Without reporting protocols, states cannot provide complete data and regions cannot review the data and ensure that it is sufficiently complete. Such incomplete data, when aggregated and compared with other state data, could lead EPA, as well as those to whom the agency reports, to draw incorrect conclusions on the status of state programs and the national program.

In addition to being incomplete, the data that EPA collects through its activity reporting forms is not comparable among states because of inconsistencies in the way that state agencies responsible for class II wells interpret the instructions on the forms. We found two key variables that states in our sample were interpreting inconsistently: (1) significant noncompliance and (2) alleged contamination of aquifers. As shown in table 4, states found different levels of significant noncompliance—those violations that, in general, pose a threat to underground sources of drinking water. For example, from fiscal year 2008 through fiscal year 2012, most states' instances of significant noncompliance represented less than 1 percent of their total well inventory. However, instances of significant noncompliance that occurred in Texas during this same period range from 2 to 11 percent of the state's total class II well inventory.

The amount of significant noncompliance reported by states can vary in part because state and EPA regional officials interpret the definition of significant noncompliance differently. Through discussions with officials in the eight states we reviewed, we discovered that at least two states use a different method than others to identify and record significant noncompliance violations. EPA guidance for determining significant noncompliance is outlined on the 7520 form, as well as through select guidance documents, and allows state discretion for determining which violations should be counted as significant noncompliance. Most states generally reported adhering to the instructions on the 7520 forms, but states' interpretation of the directions varies. For example, Texas officials reported that they consider all delinquent mechanical integrity violations as significant noncompliance, and Ohio officials said that all mechanical integrity failures are considered to be significant noncompliance regardless of their resolution. However, the guidance for reporting significant noncompliance notes that mechanical integrity violations should be counted as significant noncompliance when the loss of integrity causes the movement of fluid outside the authorized zone, if such movement may have the potential for endangering underground drinking water. Officials in Texas told us that they consider all significant noncompliance a potential threat to groundwater, but they did not indicate whether they consider proximity of underground drinking water when they determine which mechanical integrity loss violations are recorded as significant noncompliance. Ohio officials told us that significant noncompliance violations are determined based on the state's definition of significant noncompliance, even if the violation does not endanger groundwater.

Table 4. Percentage of Class II Wells with Significant Noncompliance Violations Compared to Class II Inventory in Select States for Fiscal Year 2008 through Fiscal Year 2012

	2008	2009	2010	2011	2012
California	0%	No data	0%	2%	0%
Colorado	0	0	No data	0	0
Kentucky	2	0	0	No data	No data
North Dakota	No data	0	0	0	0b
Ohio	0	No data	0	1	1
Oklahoma	0	0	0	0	0
Pennsylvania	0	0	0	No data	No data
Texas	5%	4%	2%	7%	11%

Sources: GAO analysis of EPA and state data. | GAO-14-555

Note: The amount of significant noncompliance found can vary for different reasons, including different interpretations of what is significant and inspection approaches which would make the violation rates for the different states based on different methodologies and therefore not comparable. For example, states may focus their inspections on well operators that have frequent violations, which may make the reported violation rates higher in these states than if they had inspected all of the facilities (as some states did). Moreover, it is possible for a given class II well to incur more than one citation for a significant noncompliance violation within a year. State data is provided for the federal fiscal year.

Additionally, states vary in how they interpret a second key piece of data—the instances of alleged contamination of underground sources of drinking water. The instructions included on EPA's 7520 form note that the respondent should enter the number of times a well cited for noncompliance has allegedly contaminated an underground source of drinking water. Several state agencies with whom we spoke initially had difficulty describing how they populated this field. Officials from three of the state programs with whom we spoke told us that, as they interpreted this question, they would only report instances of alleged contamination if they were confirmed as cases of contamination due to a class II well. However, officials representing the remaining five states said that they would report instances of alleged contamination without needing to confirm whether actual contamination had occurred from a class II well. Officials in California, Colorado, North Dakota, and Oklahoma indicated that they would investigate situations of potential underground drinking water contamination, and if evidence showed a reasonable suspicion or likelihood that contamination had occurred, it would be reported on the form 7520 as a case of alleged contamination. Officials from EPA Region 3, who manage the program for Pennsylvania, told us any suspected contamination of an underground source of drinking water would be reported as an instance of alleged contamination on the 7520.

EPA officials acknowledge that there may be inconsistencies in how data on these forms are reported, yet the data submitted on the 7520 forms is not subjected to any formalized review to ensure completeness or consistency. EPA regional officials said that they review the data for completeness, however, they do not have a protocol to ensure data quality or that states are reporting the data the same way in the first place. Without protocols to ensure the consistency and completeness of the data that states report, EPA regions cannot ensure the quality of the data that are being reported to headquarters and EPA may not be able to aggregate and report complete and comparable data.

EPA has also found inconsistencies in states' data that are aggregated and reported from the national UIC database, including class II data. The agency found these inconsistencies by comparing data from the simple 7520 database it created to data in the national UIC database. For example, on the basis of this data check, officials told us that they planned to refine definitions for data fields that may be yielding inconsistent results to make them more comparable across states, and they have had some initial training sessions with regions to educate them on the correct interpretation of definitions. These efforts should improve the information that is ultimately submitted in the national UIC database, according to EPA officials. These efforts could also help make the data reported on the 7520 forms more consistent and comparable; however, the agency does not plan to use the database to aggregate and report UIC program data because it plans to phase out the 7520 reports when it brings the national database online. In the meantime, however, EPA will not be able to report on the national program if it cannot aggregate the data with reasonable assurances of its completeness and consistency. Given increased public attention on the oil and gas industry, reporting on the national program is helpful for Congress, the public, and other groups to understand the program. EPA officials indicated that they are open to using the simple 7520 database for reporting if their efforts to improve the data do not distract from their efforts to improve the national UIC database.

CONCLUSION

For over 30 years, EPA and states have managed regulatory programs with safeguards that are designed to prevent contamination of underground sources of drinking water from the injection of fluids associated with oil and gas production. As domestic oil and gas production and the demand for underground injection wells continue to increase, EPA faces additional challenges maintaining sufficient oversight and enforcement of these different programs and requirements in a budget-constrained environment. States have been partners with EPA in managing their programs, yet face similar budgetary constraints. To meet its responsibilities to oversee and enforce class II program requirements, it is important that EPA ensures that state programs have information on risks to underground sources of drinking water posed by underground injection, that its oversight and enforcement are focused and efficient, and that it obtains sufficient information to monitor and report on the program nationally. We have identified several challenges that EPA faces to meet these responsibilities:

- As the class II program has developed, new risks have emerged, including overpressurization of geologic formations and potential contamination of underground sources of drinking water by diesel fuels. Without information on emerging risks on a national scale, such as through a UIC Technical Workgroup review, state class II programs may not have the information necessary to address these risks and ensure that their programs are designed to be effective at protecting underground sources of drinking water.

- EPA regions and headquarters are not consistently carrying out annual on-site evaluations of state class II programs—an activity that EPA guidance, issued in 1983 when many state programs had just been approved—identified as key to ensuring effective oversight of state programs. Limited resources have prevented EPA from conducting on-site evaluations annually. Yet, on-site evaluations may only be necessary every few years now that UIC programs have matured, and with improvements in technology, some activities that once required an on-site visit, such as file reviews, can be done remotely in some cases. Given that EPA has not conducted an overall review of the class II program since the early 1990s, the agency has not considered what activities should be priorities to ensure effective oversight of the program as it currently stands, and what resources are necessary to carry out those activities. Until EPA evaluates its guidance to determine what activities are essential for conducting effective oversight of a mature class II program, and revises its guidance as needed to reflect those activities, it cannot ensure that its oversight is effective.
- Enforcement of state and federal class II program regulations is a key EPA responsibility under the act. Yet, EPA has not taken steps to ensure that it has appropriate enforcement authority for all state program requirements. As a result, EPA cannot enforce some state program requirements. EPA's difficulty taking legal action against a violator in Illinois illustrates the importance of EPA incorporating state program requirements into federal regulations, as currently required by EPA regulations. Since the agency's 1991 rulemaking to incorporate changes to state program requirements into federal regulations, EPA has accumulated another backlog of state program requirements that need to be incorporated. Until EPA begins incorporating these changes, the backlog of state program changes it must review, approve, and incorporate into federal regulations will continue to increase. EPA officials are correct in their assessment that the process created by EPA regulations and guidance to review, approve, and separately incorporate changes in state program requirements through a rulemaking is burdensome and resource intensive. While agency officials recognize that other programs, such as the Public Water System Supervision program, have more efficient

processes for reviewing and approving state requirements, EPA has not evaluated whether it could remove the requirement for a separate rulemaking to incorporate state program changes, or somehow modify its approval process to ensure that changes to state program requirements are incorporated into federal regulations at the same time changes are reviewed and approved. Until EPA evaluates whether this requirement can be revised or eliminated to make the review, approval, and incorporation of state program changes more efficient, the process for incorporating future state program changes will remain burdensome and time-consuming.

- Congress, the public, and national groups all have great interest in the nation's oil and gas resources and their development, including how water and wastewater associated with those resources are managed. The best source of information that EPA has is the simple 7520 database, and while it has taken steps to use that database to correct information in developing and testing its national UIC database, it has not taken steps to use the 7520 database for reporting. Yet, because the 7520 data are not complete or consistent, they are not sufficiently reliable for the purposes of reporting at a national level. The same steps that the agency is taking to correct its data for the national UIC database—using consistent definitions of terms in collecting program data and having a protocol to check data quality—would help correct its 7520 data and would make that data available, perhaps earlier, to allow the agency to report on the program nationally. With some additional effort, EPA could make the 7520 database useful for reporting until the national UIC database is ready for reporting. Unless EPA takes these steps, it will be several years before EPA can provide updated information at a national level to Congress, the public, and others on the UIC program, preventing them from understanding the program and the protection being provided to underground sources of drinking water at an important juncture in the development of oil and gas in the country.

Recommendations for Executive Action

To ensure that EPA's oversight of the class II program is effective at protecting drinking water sources from the underground injection of large amounts of wastewater that will be produced with increasing domestic oil and gas production, we recommend that the Administrator of the Environmental Protection Agency take the following four actions:

- Task the UIC Technical Working Group with reviewing emerging risks, and related program safeguards, including overpressurization of formations and information on use of diesel fuels in hydraulic fracturing.
- Evaluate and revise, as needed, UIC program guidance on effective oversight to identify essential activities that EPA headquarters and regions need to conduct to effectively oversee state and EPA-managed programs.
- To ensure that EPA maintains enforcement authority of state program requirements
    - conduct a rulemaking to incorporate state program requirements, and changes to state program requirements, into federal regulations, and
    - at the same time, evaluate and consider alternative processes to more efficiently incorporate future changes to state program requirements into federal regulations without a rulemaking.
- To support nationwide reporting goals until the national UIC database is complete
    - improve the 7520 data for reporting purposes, as well as to help with quality assurance for the national UIC database, by developing and implementing a protocol for states and regions to enter data consistently and for regions to check 7520 data for consistency and completeness to ensure that data collected from state and EPA-managed class II programs are complete and comparable for purposes of reporting at a national level, and
    - in the interim, develop a method to use the 7520 database to report UIC data, including data on class II wells, until the national UIC database is fully populated with state data.

Agency Comments and Our Evaluation

We provided a draft of this report to EPA for review and comment. EPA provided written comments in which the agency expressed general agreement with the report's findings, conclusions, and recommendations. Overall, the comment letter agreed with the report's characterization of the resource challenges facing state and EPA-managed programs. EPA agreed with three recommendations, and agreed with the findings of another recommendation but did not agree with the recommended action. EPA also provided technical comments that we incorporated as appropriate. In addition, we provided appropriate sections of the draft report to the six states whose programs we reviewed. State officials from California, Colorado, North Dakota, Ohio, Oklahoma, and Texas provided technical comments, which we incorporated as appropriate.

In response to our recommendation that EPA task the UIC Technical Workgroup with reviewing emerging risks and related program safeguards, EPA agreed that to ensure effective oversight of the class II program the agency must identify emerging risks and provide state and EPA-managed class II programs with the information they need to address those risks. EPA stated that it will provide information on overpressurization to state and EPA-managed programs in the UIC Technical Workgroup's planned report on induced seismicity; and that its February 2014 guidance on class II program permitting for hydraulic fracturing with diesel fuels provides recommendations to EPA permit writers, including best practices identified by states and industry. We recognize EPA's efforts to provide additional information to state and EPA-managed programs on induced seismicity, overpressurization, and permitting hydraulic fracturing operations that use diesel fuels, but these efforts do not completely address the issues we identified. As a part of these efforts, it is important that the UIC Technical Workgroup provide information to states and EPA-managed programs on the other risks posed to underground sources of drinking water by overpressurization of formations, such as potential contamination by surface breakout of injected fluids as occurred in Oklahoma in recent years. In addition, since not all states legally require disclosure of hydraulic fracturing fluid composition through the FracFocus database, including states with large numbers of wells such as California, it is critical that the UIC Technical Workgroup provide guidance on the information that all states must collect to successfully identify and permit hydraulic fracturing operations using diesel fuels.

In response to our recommendation that EPA evaluate and revise, as needed, UIC program guidance to identify essential activities needed to effectively oversee state and EPA-managed programs, EPA stated that it generally agreed with our finding and the recommendation. The agency stated it will begin an internal dialogue among the UIC program managers at the June 2014 UIC National Managers Meeting and continue this dialogue over the next year to evaluate the effectiveness of the agency's oversight activities, and will document the results of this process and any recommendations for further action by July 2015. EPA further stated that if any change in expectations or practice is warranted, the agency will develop an advisory document that supplements its 1983 guidance on effective oversight by July 2016.

In response to our two-part recommendation that EPA conduct a rulemaking to incorporate state program requirements and changes to state program requirements into federal regulations, and that EPA evaluate and consider alternative processes to efficiently review, approve, and incorporate future changes to state program requirements into federal regulations without a rulemaking, EPA agreed with GAO's findings but it did not agree to take the recommended actions. In response to the first part of the recommendation, EPA stated that the agency cannot conduct a rulemaking to approve and codify all state program revisions without going through considerable effort to determine whether each of the changes meets the requirements of the Safe Drinking Water Act. EPA also stated that conducting a single rulemaking of this scale to incorporate all outstanding state program changes would be impractical because EPA, in conjunction with all state programs, would have to simultaneously review all state class II programs approved through both the conventional and alternative processes laid out in the Safe Drinking Water Act. EPA stated that given that the process would take many years to complete, this approach would still not ensure that all program changes are up to date in federal regulations, as other states could make changes to their programs during this time. In lieu of a single rulemaking EPA said it was conducting an ongoing process of individual rulemakings to approve and codify state program revisions in a collaborative manner with states, EPA regions, and EPA's Office of Enforcement and Compliance Assurance. However, as stated in the report, according to an analysis conducted in 2010, EPA estimated that it would take 2 to 3 years, dedicated EPA personnel, and $150,000 in outside contractor support to identify, approve, and incorporate all state program changes made since 1991 into federal regulations. By EPA's own estimate, the targeted state- by-state approach will take much longer than a single rulemaking and will face greater challenges of states continuing to make changes in the interim, leaving EPA without the ability to enforce the program, if needed. EPA provided no evidence in its comment letter that the effort it is now contemplating would be any less costly or any more efficient than the approach it assessed in 2010. For this reason, we believe that our recommendation is still necessary for EPA to carry out its responsibilities for the class II program.

In response to the second part of this recommendation, that EPA consider alternative processes to review and approve state program revisions and to incorporate state programs into federal regulations without rulemaking, EPA stated that (1) the Safe Drinking Water Act specifically requires that state UIC program revisions made in response to changes in EPA UIC regulations be approved by rule, and (2) the agency would need to revise its regulations to be able to approve state program changes without a rulemaking. We were not suggesting that EPA violate the Safe Drinking Water Act and have made changes to the recommendation language to avoid any such confusion. The focus of our recommendation is on the requirement appearing in EPA's regulation, but not in the Safe Drinking Water Act, that effectively prohibits EPA enforcement of state UIC regulations unless the latter are codified in the federal regulations. EPA stated that there are strong policy and programmatic reasons to maintain this requirement. Specifically, according to EPA, without codification it would not be possible to find a complete set of EPA approved rules for a state in one place, making it difficult for the regulated community, EPA, and the Department of Justice to determine what state program requirements are applicable and enforceable under the Safe Drinking Water Act. If EPA believes the codification process is important, it should devote the resources necessary to implementing it, or, if those resources are not forthcoming, it should consider alternative methods of maintaining federal enforceability within existing resource constraints that are likely to reduce the current backlog. In its letter, EPA states that it will continue coordination among its offices to make the program review and rulemaking process more efficient. However, it provided no detail as to what actions may produce such efficiencies, when they might take effect, or why they have not already been implemented. For these reasons, we continue to believe that EPA should explore alternative methods for ensuring the enforceability of state UIC regulations that do not rely on the rulemaking process.

In response to our recommendation that EPA develop and implement a protocol for states and regions to ensure that 7520 data are complete and comparable for purposes of reporting at the national level and that EPA develop a method to use the 7520 database to report UIC data until the national UIC database is fully populated with state data, EPA agreed with our finding and recommendation. Specifically, EPA agreed that there is room for improvement in the completeness and consistency of data submitted by the states and regions through the 7520 forms, and that the 7520 database should be completed so that it can be used as a tool to better understand national UIC activities. In addition, EPA described actions that it plans to take to implement the recommendation, including updating the instructions on the 7520 forms by adding instructions to eliminate blank fields and clarifying data definitions.

J. Alfredo Gómez
Director, Natural Resources and Environment

APPENDIX I. LIST OF STATE AND EPA-MANAGED CLASS II PROGRAMS AND WELL INVENTORY

Table 5 provides a list of EPA-managed programs, state programs with safeguards deemed effective by EPA, state programs that have adopted minimum federal underground injection control requirements, and the number of class II wells in each state in 2012.

Table 5. Class II Program Management Authority and Well Inventory

State	EPA-managed programs	State programs with safeguards deemed effective by EPA	State programs that have adopted minimum federal requirements	Number of class II wells in 2012
Alabama		X		247
Alaska		X		1,347
Arizona	X			0
Arkansas		X		1,100
California		X		49,783
Colorado		X		901
Connecticut			X	0
Delaware			X	0
Florida	X			60
Georgia			X	0
Hawaii	X	X		0
Idaho			X	0
Illinois		X		7,858
Indiana		X		1,260

Table 5. (Continued)

State	EPA-managed programs	State programs with safeguards deemed effective by EPA	State programs that have adopted minimum federal requirements	Number of class II wells in 2012
Iowa	X			7
Kansas		X		16,965
Kentucky	X			3,221
Louisiana		X		3,687
Maine			X	0
Maryland			X	0
Massachusetts			X	0
Michigan	X			1,451
Minnesota	X			0
Mississippi		X		1,212
Missouri		X		455
Montana		X		1,149
Nebraska		X		661
Nevada			X	18
New Hampshire			X	0
New Jersey			X	0
New Mexico		X		4,556
New York	X			423
North Carolina			X	0
North Dakota		X		1,290
Ohio		X		2,439
Oklahoma		X		11,134
Oregon		X		9
Pennsylvania	X			1,865
Rhode Island			X	0
South Carolina			X	0
South Dakota		X		87
Tennessee	X			19
Texas		X		52,977
Utah		X		547
Vermont			X	0
Virginia	X			12

State	EPA-managed programs	State programs with safeguards deemed effective by EPA	State programs that have adopted minimum federal requirements	Number of class II wells in 2012
Washington			X	1
West Virginia		X		710
Wisconsin			X	0
Wyoming		X		5,005

Source: EPA. | GAO-14-555.

Notes: This table excludes tribes, U.S. territories, and the District of Columbia. According to EPA, in some cases, states with 0 wells may have approval to manage their class II programs but may not have active class II programs.

APPENDIX II. KEY LEGAL PROVISIONS AND GUIDANCE GOVERNING EPA APPROVAL OF STATE PROGRAMS AND STATE PROGRAM CHANGES

This appendix presents information on provisions of law and regulations, as well as guidance that govern EPA approvals of state class II Underground Injection Control (UIC) programs.

Safe Drinking Water Act

The Safe Drinking Water Act establishes several requirements for EPA and the UIC program. These requirements include the review and approval of programs at the state level, review and approval of revisions to these programs, conditions under which EPA could manage programs for a state, and enforcement of the UIC programs. The act applies to six types of well classes, including class II wells for oil and gas related purposes.

State Program Approvals

The Safe Drinking Water Act (SDWA) of 1974 required EPA to establish the underground injection control program.[54] Among other things, EPA was to promulgate regulations specifying minimum requirements for state UIC programs. Under section 1422, each identified state was to propose its own Table 5. program, meeting the EPA requirements. EPA was then to review state submissions, and within 90 days and after reasonable opportunity for presentation of views, approve or disapprove the state program "by rule."[55] Once approved, the state has primary enforcement authority for underground water sources in the state.[56] States approved under such standards may be referred to as section 1422 states.

The act also requires states to revise their programs when EPA amends its regulation concerning requirements of state programs adding or revising a requirement.[57] In such cases, a state is to submit a notice to EPA showing that the state meets the new or revised requirement.[58] EPA is to approve or disapprove such changes using the same process as with initial program approvals, that is, by rule.[59]

Alternative Approval Process for State Class II Programs

Following EPA's June 1980 promulgation of the regulations establishing minimum requirements for state programs, Congress amended the act.[60] The December 1980 amendments created an alternative way for states to receive EPA approval of UIC class II programs in SDWA section 1425. Using this alternative, in lieu of adopting EPA's minimum requirements, a state can seek approval of its program—and primary enforcement authority—by demonstrating to EPA that its program is effective in preventing contamination of underground drinking water sources. A state must show that its program meets the same four key requirements that the EPA regulations were to address: (1) prohibition of unauthorized injections; (2) authorized and permitted injections must not endanger drinking water sources;(3) include inspection, monitoring, recordkeeping, and reporting requirements; and (4) apply to federal agencies and federal land.[61] These states may be referred to as section 1425 states.

The House Commerce Committee report for this amendment noted that 32 states already regulated underground injection related to the recovery or production of oil or natural gas and believe they have programs already in place that meet the minimum requirements of the act.[62] The report states, "it is the committee's intent that states should be able to continue these programs unencumbered with additional federal requirements if they demonstrate that they meet the requirements of the act. These requirements are the same as must be met by the administrator in establishing the regulations, thus ensuring that a state program pursuant to an alternative demonstration results in an equivalent degree of protection for drinking water sources."[63] The report also noted that, after initial approval, new demonstrations may be needed, such as if EPA revises a requirement, or "instances in which a state significantly alters a program for which a demonstration has been made, or in which the administrator determines that new information about the endangerment of drinking water supplies necessitates a new demonstration."[64] According to the report, under such circumstances, EPA would need to determine, by rule after public hearing, that a state's demonstration is no longer adequate.

Under these amendments, a state using the alternative demonstration process under section 1425 is to submit to EPA a demonstration showing its program meets the four statutory requirements.[65] The rest of the process is the same as for a regular approval: EPA is to review the state submission, and within 90 days and after reasonable opportunity for presentation of views, approve or disapprove the state program by rule. Once approved, the state has primary enforcement authority for underground water sources in the state.[66] As with the section 1422 states approved under the regular approval process, states need to take action when EPA amends its regulation adding or revising a requirement relating to class II underground injection.[67] In such cases, a state is to submit a notice to EPA demonstrating that with respect to the changed aspect of the state regulation, the state meets the four statutory requirements and represents an effective program to prevent underground injection that endangers drinking water sources.[68] EPA is to approve or disapprove such changes using the same process as with initial program approvals, that is, by rule.[69]

EPA-Managed Programs within States

The act provides that, where a state fails to propose its own program or EPA has found the state program fails to meet the minimum requirements, EPA is required to prescribe by regulation a UIC program for that state.[70] EPA can revise such a program from time to time by regulation.[71]

Enforcement of State and EPA-Managed Programs

The act authorizes EPA to enforce requirements of an applicable UIC program in a state with primary enforcement authority under certain circumstances.[72] Specifically, when EPA finds a person in violation of such a requirement, has notified the state, and after 30 days the state has not commenced appropriate enforcement action, the act requires EPA to issue an order requiring compliance or to initiate court action.[73] The act defines "applicable underground injection control program" with respect to a state with primary authority as the program or most recent amendment that has been adopted by the state and approved by EPA by rule. With respect to a state where EPA has primary enforcement authority, "applicable underground injection program" is the program which has been prescribed by the Administrator by regulation.[74]

EPA Regulations

Charged with developing the new UIC program, EPA promulgated a series of regulations in the early 1980s. EPA's regulations, as amended, are divided into distinct parts. For example, 40 C.F.R. part 144 sets forth permitting and other program requirements for all UIC programs; part 145 sets forth the requirements and procedures for approval of state programs under section 1422; and part 146 establishes technical criteria and standards for use by states and EPA in the development and implementation of state UIC programs.[75] These program regulations do not establish requirements for owners or operators of injection wells, but rather, establish requirements for state or EPA officials to use in developing UIC programs that in turn establish enforceable requirements for owners or operators of injection wells.[76] As such, there are no "background" federal regulations directly imposing requirements onto regulated parties, as is found in several other major environmental statutes.

Review and Approval of State Program Changes

In establishing the initial 1980 regulations for the UIC program, EPA interpreted the act to establish a process for changes to state programs that were originally approved under both the regular and alternative processes. As noted above, the act requires that when EPA amends the regulations to add or revise a requirement, states must in essence show they meet the requirement.[77] The act does not speak to state-initiated changes. In the regulations providing for state program approvals (part 145), EPA acknowledged that state program changes may be in response to an EPA change to a requirement, or be state-initiated.[78] EPA's regulations require states to "keep EPA fully informed of any proposed modifications to its basic statutory or regulatory authority, its forms, procedures, or priorities."[79]

Under EPA's regulations, program revisions begin when a state submits documents to EPA, as necessary under the circumstances. As to processing the revision, EPA's regulations make a distinction not found in the statute. The regulations distinguish "substantial changes" as requiring a rulemaking process, including notice in the *Federal Register*, a 30-day public comment period, and opportunity for a hearing, with notice of the approval in the Federal Register. The regulations do not require this process for nonsubstantial program changes, which may be approved by a letter from EPA to the state.[80] Since part 145 is only applicable to state programs approved under section 1422 of the Safe Drinking Water Act, this section concerning program revisions does not directly apply to the class II programs approved under section 1425 of the act. EPA officials stated, however, that the agency considers this section as guidance for 1425 states.

The regulations do not define substantial and nonsubstantial; however, EPA guidance provides a few examples of substantial changes, such as changes to a state's authority, transfer from an approved state agency to another, and changes which would make the program less stringent than applicable requirements.[81] Available documents, such as the original *Federal Register* preambles for this part, do not explain why EPA made this distinction. EPA officials said that because of the expense and time for rulemakings, EPA may have determined that such a requirement could not be intended to apply to any sort of change that could occur— such as editorial or renumbering changes—and thus carved out a less burdensome process for such changes. We note that because the regulation does not distinguish certain changes subject to specific requirements in the act—that is, state revisions that are necessitated in response to an EPA change to a requirement—and introduces a less rigorous process for some changes, it is not clear whether the regulation ensures that the statutory requirement is being met. That is, under the statute, all state revisions made in response to an EPA change are to be approved by rule. Under the regulation, however, state revisions deemed nonsubstantial can be approved by letter, rather than by rule, and it appears possible that these nonsubstantial changes could include changes made in response to an EPA requirement.

Codifying State Program Changes in Federal Regulations

In addition to the program regulations for review and approval of state programs, EPA determined in 1983 that another part was required in regulations (1) to contain EPA's requirements in states where EPA would manage the program and have primary enforcement authority and (2) to codify EPA's approval of state UIC programs.[82] On this basis, EPA promulgated part 147 in 1984.[83] The general provisions state that part 147 "sets forth the applicable Underground Injection Control (UIC) programs" in each state, and that "[r]egulatory provisions incorporated by reference (in the case of approved state programs) or promulgated by EPA (in the case of EPA-administered programs), and all permit conditions or permit denials issued pursuant to such regulations, are enforceable by the Administrator" under the act.[84] Thus, EPA established in its regulation an obligation not found in the act: that state programs must be codified into the part 147 regulations to be enforceable by EPA.[85] Available documents, such as the original Federal Register preambles for this part, do not explain why EPA added this obligation. EPA officials noted that without codification, however, there could be a due process issue whereby regulated entities may lack notice of the requirements applicable to them and enforceable by EPA.

EPA Guidance

EPA wrote guidance in 1984 to help states with the process of review and approval of state programs and state program changes, but it has not written guidance on how these changes will be codified in federal regulations under section 147. The guidance applies to all state programs approved under both sections 1422 and 1425.

The guidance, *Guidance for Review and Approval of State Underground Injection Control (UIC) Programs and Revisions to Approved State Programs*, distinguishes between substantial and nonsubstantial changes to state programs and includes examples of substantial changes. Substantial changes, including changes to a state's authority, transfer from an approved state agency to another, and changes which would make the program less stringent than applicable requirement, are to be approved by the EPA Administrator. Nonsubstantial changes may be approved by a Regional Administrator in a letter to a state's Governor.

The guidance also indicates that states will apply for EPA to review and approve their program changes and also sets out a process for EPA to review and approve the program changes in 90 days. The process involves a rulemaking procedure different from the process needed to codify the state program or program changes into section 147.

Once a state program or program change is approved, EPA's regulations indicate that it should be codified into section 147 to allow EPA to enforce the state program, if needed. However, the guidance is silent on the steps EPA headquarters and regions need to take to incorporate approved programs or program changes into section 147.

Many states had approved programs for one or more well classes by 1984, and EPA's initial promulgation of part 147 in that year codified most of them.[86] EPA typically conducted two discrete steps to approve and codify the programs: (1) approval by rule of a state program and (2) later codification of the approved program into part 147.[87] Over time, additional states obtained EPA approval for their programs. For example, Montana's class II program was approved by rule in 1996.[88] In addition, some states made changes to their programs. According to EPA headquarters officials, there have been few substantial changes to state class II programs, and EPA has not changed any requirements for class II programs that would trigger state revisions.

EPA has not, however, codified some state programs. It last conducted an exercise to update its part 147 regulations in 1991, updating the references to state regulations.[89] Even so, the revisions did not codify all of the programs approved to date. For example, West Virginia and Arkansas both had class II programs approved by EPA in 1983 and 1984, respectively, but these programs were neither included in the original part 147 nor the 1991 revisions. EPA officials could not explain why these state programs were not codified, noting that these events predated their tenure at the agency. EPA officials are aware that part 147 is out of date with respect to state program revisions, as well as these two state programs missing in their entirety. According to officials, resources have not been provided to conduct the necessary research and rulemaking. Thus, by the terms of section 147.1 in conjunction with the act, EPA is unable to enforce those state program requirements that are not contained in part 147.

APPENDIX III. OBJECTIVES, SCOPE, AND METHODOLOGY

The objectives of this report are to examine: (1) EPA and state roles, responsibilities, and resources in management the class II program; (2) safeguards select states have in place to ensure the protection of underground drinking water; (3) EPA's regulation and oversight of class II programs; and (4) the reliability of data to report on the class II program nationwide. To address all objectives, we reviewed the Safe Drinking Water Act and EPA's regulations and guidance on the Underground Injection Control (UIC) program, including class II wells. We also spoke with officials from EPA headquarters and regional offices about all aspects of the class II program. Because this report also examines states' roles in the program, we chose a sample of eight states on which to focus our analysis: California, Colorado, Kentucky, North Dakota, Ohio, Oklahoma, Pennsylvania, and Texas. These states represent a nonjudgmental sample, selected on the basis of the location of current shale oil and gas plays across the country, the number of class II wells in each state, and whether the class II program was managed by the state or EPA regions.[90] To identify current shale oil and gas plays across the country, we used Energy Information Administration regions that are organized around national shale oil and shale gas resources.[91] These regions represent diverse geography and geologic formations, as well as different oil and gas and wastewater operations. We selected at least one state in each of the six regions identified by the Energy Information Administration. We also selected states that had higher numbers of class II wells to ensure our sample represented significant class II activity. And finally, we selected states that had both state and EPA-managed programs. We spoke with state and EPA officials from programs in each of the eight states, and we visited four of these states in person. During these visits, we discussed program management and data collection and reporting with state officials, and we reviewed a sample of state inspection and violation files.

To determine the roles and responsibilities of state class II programs, we reviewed EPA regulations and guidance for the program. We also spoke with officials from six of our eight selected states about their management and reporting on the program, as well as EPA's oversight of the program.

To understand the roles and responsibilities of EPA for EPA-managed programs, we spoke with EPA headquarters and regional officials. To evaluate funding for state and EPA-managed programs, we requested class II program budget data for fiscal year 2008 through fiscal year 2012 from the six states and two EPA regional offices that manage the class II programs in the states we reviewed. We discussed budget data with state officials to assess its reliability, and we determined that it was reliable for our purpose of summarizing state program budgets. We also obtained EPA UIC grant data for federal fiscal years 2008 through fiscal year 2012, including grants provided to all UIC programs, and class II programs specifically. We interviewed EPA officials about the data, and assessed the data for any outliers or missing data, and determined that they were reliable for the purpose of reporting on class II program funding. To evaluate the resources that state programs have to administer the program, we gathered state-reported budget data on the amount of funding for the class II program supplied by the state, and compared it to annual grant allocations from EPA to states. This enabled us to develop a percentage of total class II funding that came from each state's grant compared to the funding that came from the state budget. To show the trend in funding over the last several years, we converted the EPA grant data into real fiscal year 2013 dollars. To assess staffing resources for both state and EPA-managed programs in the eight states we selected, we interviewed state and EPA regional officials about their staffing levels from fiscal year 2008 through fiscal year 2012.

To analyze and compare the safeguards that EPA and selected states have in place to ensure the protection of underground drinking water, we reviewed the basic safeguards outlined in federal regulations, state regulations, guidance, and related program documents in each of our eight selected states. To better understand the basis and purpose for regulations addressing safeguards, we reviewed a key EPA document, the *Statement of Basis and Purpose: Underground Injection Control Regulations*, and spoke with EPA officials. Based on EPA's *Statement of Basis and Purpose*, we identified seven program safeguards required for EPA-managed programs. To compare the safeguards in each state and EPA-managed program in eight selected states, we developed a table for each safeguard that categorized and compared the safeguards used by the programs in each of the eight states. We identified state program safeguards in state regulations, program guidance, and other documents. Our review included a summary and comparison of the regulations and guidance that establish state and EPA-managed program safeguards, but we did not analyze the technical sufficiency of those safeguards. Additionally, we gathered data from the states on inspections and alleged contamination of underground sources of drinking water. We discussed these data with state officials to assess their reliability for reporting, and in some instances, used limited EPA data to corroborate the state-reported data. We determined that these data were reliable to report for our purposes of describing state program safeguards. To identify any potential gaps in the safeguards in place, we obtained and analyzed EPA reviews such as the Mid-Course Review and an EPA contracted report by the technical and strategic consulting group, Cadmus.[92] We also spoke with EPA and state officials about potential weaknesses in the safeguards, the effectiveness of the safeguards, and any issues that may affect the protectiveness of the safeguards.

To examine EPA's regulation and oversight of the class II program, we focused on the regulation and oversight of programs in our eight selected states: six state programs and two EPA-managed programs. To understand EPA's regulation of the programs in the eight states, we reviewed EPA's regulations for state programs. A key aspect of these regulations is the need to incorporate state regulations and revisions to these state regulations into federal regulations; this allows EPA to enforce state programs, when appropriate. To assess the extent to which EPA had accomplished the incorporation of state regulations into federal regulations, we identified situations in which the states we reviewed had updated or changed regulations and discussed the status of EPA's approval and incorporation of these changes. We also reviewed EPA's guidance and processes for requesting and approving aquifer exemptions, and we requested documentation of exemptions from EPA regions and headquarters. To understand EPA's oversight responsibilities and evaluate the extent to which they are being carried out, we reviewed EPA guidance outlining effective oversight of state programs. We spoke with EPA headquarters and regional officials regarding how they completed the actions outlined in the guidance. We also obtained and reviewed the annual reports that EPA regions are to write for state programs, Memoranda of Agreements between EPA and each state program, and relevant grant reports from states.

Finally, in order to examine the reliability of the data that EPA has on class II programs, we interviewed EPA headquarters' officials about their various data collection methods and reviewed related documentation and guidance, as available. We identified three methods that EPA headquarters uses to collect data on the UIC program, including the class II program, and discussed the data collected by each method with EPA officials. We determined that one source of data contained the information we needed—the state activity reporting forms (7520 forms) that each state completes twice per year. We requested these forms from EPA for fiscal year 2008 through fiscal year 2012 and obtained all the forms that EPA said it had. The forms were in paper format, which we saved and transferred into an electronic database to analyze. To assess the reliability of the data for our reporting purposes, we took steps to evaluate the completeness and comparability of the data. To evaluate the completeness of the data, we performed basic checks of the data supplied by EPA to identify missing data. We found numerous missing data that resulted from incomplete or missing forms and mislabeled forms. We interviewed headquarters and regional EPA officials about this data, and they said that they had provided all the forms available to them. We also performed basic data checks on the consistency of the data provided by discussing state data with officials in the states we visited. We interviewed state officials about how they completed the 7520 forms and what data they used to complete the forms, and we discussed their understanding of the instructions on the 7520 form. From our evaluation of the completeness and consistency of the data, we determined that the 7520 data were not sufficiently complete or consistent, and we decided not to report the data, with two exceptions: we reported the number of instances alleged contamination reported by each of the eight states we reviewed because these data were solely descriptive, and we reported data on significant violations reported by the eight states to show the inconsistency in state reporting. In addition to discussing the 7520 data, we spoke with state and regional officials about the national UIC database and its development. We also reviewed documentation related to the national UIC database in development, including its data dictionary and business rules to example the progress and status of the database.

We conducted this performance audit from November 2012 to June 2014 in accordance with generally accepted government auditing standards. Those standards require that we plan and perform the audit to obtain sufficient, appropriate evidence to provide a reasonable basis for our findings and conclusions based on our audit objectives. We believe the evidence provides a reasonable basis for our findings and conclusions based on our audit objectives.

APPENDIX IV. SELECT STATES' CLASS II PROGRAM SAFEGUARDS AND RELATED REGULATIONS

In the 1980s, to fulfill its statutory obligations under the Safe Drinking Water Act, EPA developed safeguards to protect underground drinking water sources from contamination by fluids that are injected into underground formations and could leak into other formations that contain sources of drinking water. Specifically, the safeguards help ensure that wells are designed to prevent fluids that are injected into underground formations from endangering underground drinking water sources. This appendix and the following tables summarize requirements for key class II program safeguards for the programs we reviewed in eight states: California, Colorado, Kentucky, North Dakota, Ohio, Oklahoma, Pennsylvania, and Texas. The programs in two of these states— Kentucky and Pennsylvania—are managed by EPA regions under EPA regulations, while the remaining six programs are state programs approved by EPA.

Area of Review

To prevent fluids from entering an underground drinking water source, state UIC program directors determine an area of review necessary to obtain a permit for new injection wells. This is an area around an injection well where pressure in the injection zone may cause the migration of fluids into an underground source of drinking water, according to EPA documents. Before fluids can be injected into a new well, the state director must consider information on other active or abandoned wells in the area, and the corrective action status of any defective wells. The size of the area of a review can be a fixed radial distance of one-quarter mile or greater, or it can be calculated by a formula that considers the injection rate, movement of fluids through the injection zone, and the size of the injection zone among other factors. Table 6 shows the requirements for area of review for the eight programs we reviewed.

Geologic Information, Water-Bearing Formations, and Confining Layers

To ensure that fluids do not travel through weak areas of a confining layer, EPA regulations require that appropriate geological data on the injection zone and confining zone be considered before issuance of a permit to inject fluids. Programs can rely on existing data, but a permit cannot be issued until confining formations are determined to be sound and capable of containing injected fluids. Table 7 shows the requirements for geologic and other information for the eight programs we reviewed.

Table 6. Select State Class II Requirements for the Area of Review for New Injection Wells

State	Area of review (AOR)	Exceptions	Types of wells or holes to be identified in AOR	Corrective action requirements
California	AOR is not defined in state law or regulation[a]	None identified	All injection, production, and plugged and abandoned wells.	No corrective action requirements per se, but operators must prove that plugged and abandoned wells will not have an adverse effect on the project or cause damage to life, health, property, or natural resources.
Colorado	¼- to ½ mile	None identified	Underground disposal: oil and gas wells, domestic and irrigation water wells of public record within ¼ mile; oil and gas wells currently producing from the injection zone within ½ mile.[b] Enhanced recovery and liquid hydrocarbon storage: wells, including drilling wells and dry and abandoned wells, within ¼ mile.	Underground disposal: remedial action is required for any well within 1/4 mile of the proposed well(s) in which the injection zone is not adequately confined. Enhanced recovery and liquid hydrocarbon storage: a plan for performance of needed remedial action on wells penetrating the injection zone within ¼ mile is required.

Table 6. (Continued)

State	Area of review (AOR)	Exceptions	Types of wells or holes to be identified in AOR	Corrective action requirements
Kentucky[c]	Either (1) the distance to which migration of injected fluid into an underground source of drinking water could be expected, calculated based on factors detailed in the regulation, or (2) ¼ mile	None identified	All known wells (producing wells, injection wells, abandoned wells, dry holes, and water wells).	Corrective action is required for all wells which are improperly sealed, completed or abandoned and which penetrate (1) the injection zone or (2) a formation affected by injection pressure.
North Dakota	¼ mile	None identified	Injection wells, producing wells, plugged wells, abandoned wells, drilling wells, dry holes, and water wells.	Corrective action is required for all wells penetrating the injection zone if deemed necessary.
Ohio	Either (1) ¼ mile for wells injecting up to 200 barrels per day or (2) ½ mile for wells injecting more than 200 barrels a day	An alternate AOR can be designated for good cause shown.	All wells penetrating the formation proposed for injection.	Corrective action is required to prevent fluid from moving into a source of underground water.
Oklahoma	¼ mile for noncommercial disposal wells; ½ mile for commercial disposal wells 0 miles for passive injection into simultaneous injection wells	None identified	Simultaneous injection wells: unplugged or mud-plugged boreholes Disposal wells: abandoned, producing or drilling wells and dry holes	None identified

State	Area of review (AOR)	Exceptions	Types of wells or holes to be identified in AOR	Corrective action requirements
	¼ mile for active injection into simultaneous injection wells, which are wells that inject or dispose of salt water at the same time they are producing oil and/or gas to the surface[d]			
Pennsylvania[c]	Either (1) the distance to which migration of injected fluid into an underground source of drinking water could be expected, calculated based on factors detailed in the regulation, or (2) ¼ mile	None identified	All known wells (producing wells, injection wells, abandoned wells, dry holes, and water wells)	Corrective action is required for all wells which are improperly sealed, completed or abandoned and which penetrate (1) the injection zone or (2) a formation affected by injection pressure.
Texas	¼ mile for disposal wells[e]	A variance can be granted upon proof that it will not result in a material increase in the risk of fluid movement into freshwater strata or to the surface.	Wells that penetrate the proposed disposal zone	None identified[f]

Sources: GAO analysis of federal and state regulations. | GAO-14-555

[a] According to the state program's website, regulators make reference to federal UIC regulations in defining the "area affected by the project." Accordingly, the distance is either (1) that to which migration of injected fluid into an underground source of drinking water could be expected, calculated based on factors detailed in federal regulation, or(2) ¼ mile.

[b] According to Colorado officials, this also includes plugged and abandoned, and dry and abandoned oil and gas wells.

[c] EPA implements the Underground Injection Control program in Kentucky and Pennsylvania, so this table shows federal requirements applicable in those states.

[d] If injection is by gravity flow, no AOR is required. If injection is by positive pump pressure, a ¼ mile AOR is required. According to Oklahoma state officials, however, every injection well permit includes a condition that states that "Disposal is prohibited if operation of the well for disposal pollutes or endangers subsurface treatable waters."

[e] According to Texas officials, Texas can require a larger area of review in areas when necessary.

[f] According to Texas state officials, every injection well permit includes a condition that states that "should it be determined that such injection fluid is not confined to the approved interval, then the permission given herein is suspended and the disposal operation must be stopped until the fluid migration from such interval is eliminated." Officials also told us that the state recommends corrective action to reenter and properly plug area wells if evidence is presented to demonstrate that they are a potential conduit for contamination.

Table 7. Select Information Collected by States Regarding Geologic Information, Water-bearing Formations, and Confining

State	Injection zone information	Base of water-bearingformations	Presence of confining layer above injection zone
California	Intended zone for completion (injection) and study of porosity, permeability, average thickness, extent, fracture gradient, temperature and pressure, and original and residual oil, gas, and water saturations	Estimated depth of base of freshwater	Not discussed

State	Injection zone information	Base of water-bearingformations	Presence of confining layer above injection zone
Colorado	Formation name, total dissolved solids, fracture gradient, porosity, and permeability	Not discussed	Not discussed in permitting documents; however, according to Colorado officials, a confining layer is defined during the engineering review.
Kentucky	Appropriate geologic information on injection zone, including lithologic (physical description), geologic name, thickness, depth, and fracture pressure	Depth to bottom of all underground sources of drinking water that may be affected by the injection	Appropriate information on confining zone, including lithologic description, geological name, thickness, depth, and fracture pressure.
North Dakota	Geological name and depth of injection zone and lithologic description	Geologic name of lowest known freshwater zone and depth of base of freshwater zone	Geologic name of confining zone and its thickness. All new injection wells must be sited so they inject into a formation that has confining zones that are free from known open faults or fractures within the area of review.
Ohio	Proposed geologic formation for injection	Record of formation and any show of oil, gas, freshwater, or brine According to Ohio officials, deepest underground source of drinking water is identified, and casing is set at least 50 feet below it	Not discussed in permitting documents; however, Ohio officials told us that a confining zone must be present.

Table 7. (Continued)

State	Injection zone information	Base of water-bearingformations	Presence of confining layer above injection zone
Oklahoma	Target formations and depths for	Base of treatable water	Not discussed in permitting documents; however, according to Oklahoma officials, when the fluid injection rate is 1,000 barrels per day or less, or equivalent rate for any fraction of 24 hours, an overlying strata of at least 200 feet in thickness between the lowest base of freshwater and the top of the proposed interval of injection is considered sufficient evidence of freshwater protection. When the fluid injection rate is greater than 1,000 barrels per day or equivalent rate for any fraction of 24 hours, an overlying strata of at least 500 feet in thickness between the lowest base of freshwater and the top of the proposed interval of injection is considered sufficient evidence of freshwater protection. If the overlying strata is less than the requirements outlined above, the Commission may administratively approve injection provided a finding is

State	Injection zone information	Base of water-bearingformations	Presence of confining layer above injection zone
			made that such injection will not initiate fractures through the overlying strata into the freshwater strata.
Pennsyl-vania	Appropriate geologic information on injection zone, including lithologic, geological name, thickness, depth, and fracture pressure	Depth to bottom of all underground sources of drinking water that may be affected by the injection	Appropriate information on confining zone, including lithologic description, geological name, thickness, depth, and fracture pressure.
Texas	For disposal wells: name of disposal formation For enhanced recovery wells: name of formation, lithology, average thickness, pressure, porosity, permeability	Base of usable quality water and base of the deepest underground source of drinking water for both disposal and enhanced recovery wells	Not discussed in permitting documents, however, Texas officials told us that evidence of confining layers are determined by the agency's Groundwater Advisory Unit. This organization then drafts a "no-harm" letter, which is required in order to complete a disposal well permit application.

Sources: GAO analysis of federal and state class II permit forms. | GAO-14-555

Casing, Cementing, Tubing, and Packer

To prevent fluid from moving through a faulty injection casing, EPA set specific requirements for well construction. According to EPA's *Basis and Purpose* report, some wells only need surface casing, or casing that extends the length of the bore through the formation in which shallow drinking water exists, while other wells may need multiple sets, or "strings," of casing depending on the depth of the well and the surrounding geologic formation.[93] For example, intermediate casing can be necessary to protect other underground resources such as coal beds or gas storage zones. The deepest layer of casing, known as the long- string or production casing, isolates injected fluids into designated formations. EPA also requires that each layer of casing be surrounded by cement and suggests that specific equipment called tubing and packer be used. Tubing is typically steel or plastic pipe inserted inside the production casing, which isolates the casing from the fluid injected into the well. Tubing set on a packer allows well operators to directly inject fluids into formations and prevents corrosion by not allowing injected fluids to contact the casing wall. A packer is a mechanical device that sits below the tubing and locks into the casing wall, sealing the space between the tubing and casing, called the annulus, from the injection zone. Table 8 shows requirements for casing and cementing for the eight programs we reviewed.

Mechanical Integrity

To prevent fluids from leaking out or up through the wellbore, along the outside of the cement surrounding the casing, EPA's regulations require that a well needs to demonstrate mechanical integrity, or the absence of leaks. Mechanical integrity testing involves, for example, increasing the pressure in the tubing and ensuring that the well is able to hold that pressure for a period of time. Verification first occurs prior to the well being authorized as ready for injection, with subsequent verification occurring at least once every 5 years during the operation of most wells. Table 9 shows the requirements for mechanical integrity testing for the eight programs we reviewed.

Table 8. Select State Requirements for Casing and Cementing of New Class II Wells

State	General surface casing andcementing	General intermediate casing and cementing	General production casing and cementing
California	Surface casing is required. The surface casing must be cemented into or through a competent bed[a] at a depth that will allow complete well shut-in without fracturing the formation immediately below the casing shoe.	Intermediate casing may be required to protect oil, gas, and freshwater zones and to seal off anomalous pressure zones, lost circulation zones, and other drilling hazards. Intermediate and production casings must be cemented to at least 500 feet above oil and gas zones and anomalous pressure intervals and at least 100 feet above the base of the freshwater zone.	Production casing is required. The production casing must be cemented. At least 100 feet of overlap between the production string and the next larger casing string shall be required and the overlap cemented. Intermediate and production casings must be cemented to at least 500 feet above oil and gas zones and anomalous pressure intervals and at least 100 feet above the base of the freshwater zone.
Colorado	Surface casing is required. When subsurface conditions are unknown, the surface casing must be run to reach below all known or reasonably estimated usable domestic freshwater levels; and prevent blowouts or uncontrolled flows and must be cemented so as to fill the annulus to the top of the hole.	Intermediate and production casing are not specifically required. When used, cement shall be pumped behind the intermediate and/or production casing to at least 200 feet above the top of the shallowest known production horizon. All freshwater aquifers that are exposed below the surface casing shall be cemented behind the	Intermediate and production casing are not specifically required. When used, cement shall be pumped behind the intermediate and/or production casing to at least 200 feet above the top of the shallowest known production horizon. All freshwater aquifers that are exposed below the surface casing shall be cemented behind the

Table 8. (Continued)

State	General surface casing andcementing	General intermediate casing and cementing	General production casing and cementing
	When subsurface conditions are known, surface casing must be set and cemented to the surface at a depth and in a manner sufficient to protect freshwater and ensure against blowouts or uncontrolled flows.	production casing. All such cementing around an aquifer shall consist of a continuous cement column extending from at least 50 below the bottom of the freshwater aquifer that is being protected to at least 50 feet above the top of said freshwater aquifer. When freshwater aquifers are at such depth as to make it impractical or uneconomical to cover or isolate all freshwater aquifers, the intermediate and/or production string can be stage cemented.	production casing. All such cementing around an aquifer shall consist of a continuous cement column extending from at least 50 below the bottom of the freshwater aquifer that is being protected to at least 50 feet above the top of said freshwater aquifer. When freshwater aquifers are at such depth as to make it impractical or uneconomical to cover or isolate all freshwater aquifers, the intermediate and/or production string can be stage cemented.
Kentucky	Casing and cementing requirements are determined based on geological factors	Casing and cementing requirements are determined based on geological factors.[b]	Casing and cementing requirements are determined based on geological factors.[c]
North Dakota	Surface casing is required. The surface casing must be set and cemented at least 50 feet below the base of the Fox Hills formation and must fill the annular space behind the casing to	Intermediate casing is required and must be cemented.	Production casing is required and must be cemented.

State	General surface casing andcementing	General intermediate casing and cementing	General production casing and cementing
	the bottom of the cellar, if any, or to the surface of the ground.		
Ohio	Surface casing must be set and cemented to at least 50 feet below the base of certain underground sources of water[d] or at least 50 feet into competent bedrock, whichever is deeper. Surface casing shall be cemented before drilling through hydrocarbon bearing flow zones or zones that contain concentrations of total dissolved solids exceeding 10,000 milligrams per liter (mg/l) unless otherwise approved. Sufficient cement shall be used to fill the annular space outside the casing from the seat to the ground surface or to the bottom of the cellar.	Intermediate casing must be set and cemented in a competent formation if groundwater containing less than 10,000 mg/l of total dissolved solids (TDS) is encountered below the base of the surface casing; the well runs through a gas storage reservoir within the storage protective boundary; drilling through particular formations; or in other situations as determined by the chief of the oil and gas division. For each permanently set intermediate string of casing, tail cement shall extend from the seat to a point at least 500 true vertical feet above the casing seat, or to a point at least 200 feet above the seat of the next larger diameter casing string. If the intermediate wellbore penetrates flow zones, cement	Cement production casing no less than 300 feet above the top of the injection zone.

Table 8. (Continued)

State	General surface casing andcementing	General intermediate casing and cementing	General production casing and cementing
		must be placed at least 500 feet above the uppermost flow zone.[e]	
Pennsylvania	Surface casing must be installed from the surface to at least 50 below the base of the lowermost underground source of drinking water. The casing is to be cemented by recirculating cement to the surface or using no less than 120 percent of calculated annual volume.	No specific requirements; casing and cementing requirements determined based on geological factors.	For enhanced recovery wells, install tubing or long string casing to the injection zone. For saltwater injection wells, install tubing and long string casing to the injection zone. Place sufficient cement to fill the calculated volume to a point 50 feet above the injection zone.
Oklahoma	Casing must be run and cemented from bottom to top with a minimum depth that is the greater of 90 feet below the surface or 50 feet below the base of treatable water. For commercial saltwater disposal wells, surface casing must be set and cemented at least 50 feet below the base of the treatable water bearing zone.	Any casing run in addition to the surface casing must be cemented from the base of the casing string to a minimum height of the greater of 5 percent of the depth to which the string is set or a height of 200 feet.	Any casing run in addition to the surface casing must be cemented from the baseof the casing string to a minimum height of the greater of 5 percent of the depth to which the string is set or a height of 200 feet. For commercial saltwater disposal wells, the production casing must be set and cemented through the injection zone to a height of at least 250 feet above the disposal zone.

State	General surface casing andcementing	General intermediate casing and cementing	General production casing and cementing
Texas	Surface casing must be set and cemented to protect all usable-quality water strata. The Groundwater Advisory Unit of the Oil and Gas Division will set the protection depth no deeper than 200 feet below the specified depth.[f]	Intermediate casing strings must be cemented from the shoe to a point at least 600 feet above the shoe. If a productive or potential flow zone, or a zone with corrosive formation fluids is open to the wellbore above the casing shoe, the casing shall be cemented from 100 to 600 feet above the zone depending on how the top of the cement is determined, or at least 200 feet above the shoe of the next shallower casing string that was set and cemented in the well (or to surface if the shoe is less than 200 feet from the surface).	Production casing strings must be cemented to the surface or at least 600 feet above the shoe. If any productive zone, potential flow zone and/or zone with corrosive formation fluids is open to the wellbore above the casing shoe, the casing shall be cemented in a mannerthat effectively seals off all such zones as specified for intermediate casing. In the event that the distance from the casing shoe to the top of the shallowest productive zone, potential flow zone and/or zone with corrosive formation fluids make required cementing impossible or impractical, the multistage process may be used to cement the casing in a manner that will effectively seal off all such zones, and prevent fluid migration to or from such zones within the wellbore.

Sources: GAO analysis of federal and state regulations. | GAO-14-555

[a] Bed competence is generally defined as a relative value that refers to the ability of the bed to lift itself without interior adjustment or crumpling.

[b] According to EPA regional officials, any casing run in addition to the surface casing must be cemented from the base of the casing string to the surface. If the intermediate casing string does not extend to the surface, it must overlap the next larger casing and have adequate cement behind the casing to prevent the movement of fluids into or between any underground sources of drinking water.

[c] According to EPA regional officials, production casing must be cemented into the confining zone to isolate the injection zone from any underground sources of drinking water

[d] For saltwater injection wells, casing must be set below the deepest underground source of water containing less than 10,000 mg/l chlorides. For enhanced recovery projects, casing must be set below the deepest underground source of water containing less than 10,000 mg/l total dissolved solids or less than 5,000 mg/l chlorides. For wells generally, the casing must be set below the deepest underground source of drinking water. There are alternative requirements for areas where bedrock underground sources of drinking water cannot be mapped.

[e] Intermediate casing may generally be set at the discretion of the owner to isolate flow zones, lost circulation zones, or other geologic hazards.

[f] Wells drilled to 1,000 feet or less do not require surface casing when no shallow gas sands or abnormally high pressures are known to exist at depths shallower than 1,000 feet, and the production casing is cemented from the shoe to the ground surface by the pump and plug method.

Table 9. Select State Requirements for Mechanical Integrity Testing of Class II Underground Injection Control (UIC) Wells

State	Frequency of mechanical integrity testing	Required action in response to mechanical integrity test failure	Estimated percent of tests observed by program officials in 2012
California	California regulations require a two-part test to determine mechanical integrity: (1) A test to check for leaks in the casing-tubing annulus must be conducted prior to commencing	None identified	30%

State	Frequency of mechanical integrity testing	Required action in response to mechanical integrity test failure	Estimated percent of tests observed by program officials in 2012
	injection operations and at least once every 5 years, or at the request of Division. (2) A test to determine that there is no fluid migration behind the casing, tubing, or packer must be performed: • within 3 months after injection has commenced, • at least once a year for water disposal wells, • at least every 2 years for waterflood wells, • at least every 5 years for steamflood wells, • following any anomalous rate orpressure change and, • at other times, at the request of the Division.		
Colorado	Dedicated injection wells must undergo a two-part mechanical integrity test before any fluids are injected into the well and at least once every 5 years. Simultaneous injection wells are required to undergo a two-part mechanical integrity test before any fluids are injected into the well.	Wells that lack mechanical integrity must be shut-in immediately and repaired or plugged and abandoned within 6 months of failing the test.	100
Kentucky	Every 5 years during the life of the well.	The EPA UIC program director provides written notice of failure of mechanical integrity to the owner/operator; the owner/operator is generally required to cease injection within 48 hours of receipt of	100

Table 9. (Continued)

State	Frequency of mechanical integrity testing	Required action in response to mechanical integrity test failure	Estimated percent of tests observed by program officials in 2012
		injection interval. The operator must submit proof of mechanical the notification letter. The owner/operator must either plug or remediate the well; the owner/operator may not resume injection until receiving written approval from EPA.	
North Dakota	All injection wells must demonstrate continual mechanical integrity, and undergo mechanical integrity testing prior to commencing operations and at least once every 5 years.	If any monitoring indicates the movement of injection or formation fluids into under-ground sources of drinking water, the state shall prescribe such additional requirements for construction, corrective action, operation, monitoring, or reporting	100

State	Frequency of mechanical integrity testing	Required action in response to mechanical integrity test failure	Estimated percent of tests observed by program officials in 2012
		as are necessary to prevent such movement.	
Ohio	If monthly monitoring for leaks in saltwater disposal wells and enhanced recovery wells is not feasible, a mechanical integrity test shall be performed once every 5 years.	If mechanical failures or downhole problems cause contamination of the land, surface waters, or subsurface waters, the owner must cease operation immediately until the state determines problems have been corrected.	100
Oklahoma	For commercial disposal wells, a mechanical integrity test must be performed prior to commencement of operations and then every year. For other enhanced recovery and disposal wells, a mechanical integrity test must be performed prior to commencement of operations and then every 5 years, and when down-hole equipment is moved or replaced. Alternatively, monthly monitoring of injection rate, volume, and casing tubing annulus pressure can be conducted. For simultaneous injection wells, reports on mechanical integrity must be filed annually.	If a mechanical integrity problem occurs, the operator must notify the field inspector within 24 hours of its discovery and submit a notice of the failure and plan to repair and/or retest the well within 5 days. The state may shut down the well if the problem indicates that injected substances are	92

Table 9. (Continued)

State	Frequency of mechanical integrity testing	Required action in response to mechanical integrity test failure	Estimated percent of tests observed by program officials in 2012
		not or may not be entering the authorized integrity and that injected substances are going into and are confined to the permitted injection interval, at that point, the Manager of UIC may authorize recommencement of injection.	
Pennsylvania	Every 5 years during the life of the well.	The EPA UIC program director provides written notice of failure of mechanical integrity to the owner/operator; the owner/operator is generally required to cease injection within 48 hours of receipt of the notification letter. The owner/operator must	79

State	Frequency of mechanical integrity testing	Required action in response to mechanical integrity test failure	Estimated percent of tests observed by program officials in 2012
		either plug or remediate the well; the owner/operator may not resume injection until receiving written approval from EPA.[a]	
Texas	Disposal and enhanced recovery wells completed with surface casing and cemented through the entire interval of protected usable-quality water shall be tested for mechanical integrity at least once every 5 years and after every workover. Other disposal wells shall be tested at the frequency prescribed by their permits.	None identified[b]	31

Sources: GAO analysis of federal and state regulations.| GAO-14-555.

[a] According to EPA regional officials, operators are given 90 days to return to compliance, and violations are elevated to the level of significant noncompliance if not resolved within 180 days.

[b] According to a state manual, a well that fails a mechanical integrity test must be shut in immediately, repaired, and successfully retested or plugged. Typically, the operator will be required to repair and retest the well within 60 days.

Injection Pressure

Another measure to ensure that fluids do not travel through the confining layers and into a source of drinking water is to control the pressure at which fluids are injected, or injection pressure. EPA regulations require that well injection pressure should be controlled to avoid initiating fractures or propagating existing fractures in the confining zone adjacent to underground sources of drinking water. Table 10 shows the requirements for injection pressure for the eight programs we reviewed.

Table 10. Select State Requirements for Maximum Allowable Injection Pressure or Injection Rate

State	Allow injection pressure above fracture pressure	Information on injection pressure or injection rate limits collected by the state
California	No[a]	Maximum anticipated surface injection pressure (pump pressure) and daily rate of injection by well. To determine the maximum allowable surface injection pressure, a step-rate test shall be conducted prior to sustained liquid injection unless it is determined that injection pressure will be maintained considerably below the estimated pressure required to fracture the zone of injection.
Colorado	No[b]	Minimum and maximum fluid injection rates and pressures.
Kentucky	No[c]	Average and maximum daily volume and injection pressure.
North Dakota	Probably not[d]	Estimated average injection rate and pressure, and estimated maximum injection rate and pressure.
Ohio	No	Estimated average and maximum amount of brine to be injected daily into the proposed injection well, and the estimated average and maximum pressure to be used for injecting brine into the proposed injection well. Maximum allowable surface injection volume by formula.

State	Allow injection pressure above fracture pressure	Information on injection pressure or injection rate limits collected by the state
Oklahoma	No[e]	Requested injection rate and pressure.
Pennsylvania	No	Average and maximum daily volume and injection pressure.
Texas	None identified[g]	Maximum daily volume and injection pressure.

Sources: GAO analysis of state class II permit application forms. | GAO-14-555

[a] According to California officials, in practice, California allows injection above fracture pressure in certain specific cases. Diatomite rock requires steam injection over fracture pressure to create permeability channels in the formation, so that the oil can flow and be extracted. California plans to address injection above fracture pressure in future rulemakings.

[b] Maximum injection pressure will be set upon approval of a permit application, and down-hole pump configurations shall be designed to inject below the injection zone fracture gradient.

[c] Injection pressure shall not exceed maximum calculated to prevent new or propagation of fractures in the confining zone and shall not cause movement of injection or formation fluids into a protected aquifer.

[d] Except during stimulation, injection pressure shall not exceed 90% of the fracture pressure of the injection zone to assure that the pressure in the injection zone does not initiate new fracture or propagate existing fractures in the injection zone. The injection pressure shall not initiate fractures in the confining zone or cause the movement of injection or formation fluids into an underground source of drinking water.

[e] For commercial disposal wells, the injection pressure must not approach or exceed the demonstrated fracture gradient of the injection zone. For enhanced recovery and other disposal wells, applicants must demonstrate that injection into the proposed zone will not initiate fractures through the overlying strata that could enable the injection fluid or formation fluid to enter freshwater strata.

[f] Injection pressure shall not exceed maximum calculated to prevent new or propagation of fractures in the confining zone and shall not cause movement of injection or formation fluids into a protected aquifer.

[g] According to a state manual, however, pressure limitations are established in permits that provide adequate assurance that injection will not initiate fractures in the confining zones. No injection or disposal well will be permitted where faults, fractures, structures, or other geologic factors indicate that isolation of the authorized injection or disposal zone is jeopardized. Pressure limitations are established to provide adequate assurance that injection will not initiate fractures in the confining zones.

Plugging and Abandonment

To prevent fluid from moving through improperly abandoned wells, EPA regulations require that after operation of a well ceases the wellbore be plugged with cement. Table 11 shows the requirements for plugging and abandoning new wells for the eight programs we reviewed.

Table 11. Select State Requirements Regarding Well Abandonment and Plugging for Class II Wells

State	Plugging plan /notification required inadvance of well plugging	Requirement that plugging be witnessed orreported	Bond / financial assurance requirement
California	Yes	Yes	Operators of active wells must file individual or blanket bonds. For class II commercial wastewater disposal wells, the individual bond requirement is $100,000. For other active wells, the individual bond requirement is $25,000 for wells less than 10,000 feet deep and $40,000 for wells 10,000 or more feet deep. Operators of 20 or more active wells of any type can file a blanket bond of $200,000 for up to 50 wells or $400,000 for more than 50 wells. Operators of idle wells must pay annual fees, provide surety, or file individual or blanket bonds. For idle wells, the individual bond requirement is $5,000. Operators may also file a blanket bond in the amount of $2,000,000 to cover both active and inactive wells of any type.
Colorado	Yes	Yes	Operators of class II commercial Underground Injection Control (UIC) wells must provide financial assurance of either $50,000 or another amount as agreed to with or determined by the state. Operators of other types of wells must provide either individual financial assurance

State	Plugging plan /notification required inadvance of well plugging	Requirement that plugging be witnessed orreported	Bond / financial assurance requirement
			of $10,000 for wells less than 3,000 feet deep and $20,000 for wells 3,000 or more feet deep, or blanket financial assurance of $60,000 for less than 100 wells or $100,000 for 100 or more wells. Financial assurances must be increased if an operator has "excess inactive wells."[a]
Kentucky	Yes	Yes	Permittees must maintain financial resources to close, plug, and abandon the well in a manner prescribed by the EPA Regional Administrator until the well has been plugged and abandoned or converted or the permit has been transferred to a new permittee who has demonstrated financial responsibility. [b]
North Dakota	Yes	No c	Operators must submit individual, blanket, or unit bonds, or alternate approved sureties. The individual bond amount is generally $50,000 per well, though a lesser amount may be approved for noncom-ercial disposal wells less than 2,000 feet deep. The blanket bond amount is generally $100,000 for up to six wells, but commercial disposal wells are not eligible for blanket bonds. The unit bond requirement shall be specified by the state for areas under unitized management.
Ohio	Yes	Yes	Operators must provide bonds in the amount of $5,000 for a single well or $15,000 for all of an operator's wells.
Oklahoma	Yes	Yes	Operators must provide surety (in the form of a bond, letter of credit, or other negotiable instrument) equal to the total estimated cost of plugging and abandonment, up to $25,000.

Table 11. (Continued)

State	Plugging plan /notification required inadvance of well plugging	Requirement that plugging be witnessed orreported	Bond / financial assurance requirement
Pennsylvania	Yes	Yes	Permittees must maintain financial resources to close, plug, and abandon the well in a manner prescribed by the EPA Regional Administrator until the well has been plugged and abandoned or converted or the permit has been transferred to a new permittee who has demonstrated financial responsibility. [d]
Texas	Yes	Yes	A bond of at least $25,000 is required for 1 to 10 wells, $50,000 for 11 to 99 wells, and $250,000 for more than 100 wells.

Sources: GAO analysis of federal and state information. | GAO-14-555

[a] An operator has "excess inactive wells" if the operator's inactive well count exceeds such operator's financial assurance amount divided by $10,000 for inactive wells less than 3,000 feet deep or $20,000 for inactive wells 3,000 or more feet deep. For each excess inactive well, the financial assurance must be increased by $10,000 for wells less than 3,000 feet deep or $20,000 for inactive wells greater than or equal to 3,000 feet deep, though this requirement may be modified upon approval of a plan for reducing inactive wells.

[b] In Kentucky, financial responsibility requirements vary based on well depth and casing: costs range from $2,300 for a well less than 500 feet deep with casing cemented to the surface to $7,400 for a well deeper than 2,000 feet with casing cemented to a point below the surface.

[c] According to North Dakota officials, although state regulations do not require that state officials witness well plugging, in practice, North Dakota officials witness all well pluggings.

[d] In Pennsylvania, according to EPA officials, a third-party bid for the cost of plugging serves as the basis for the financial assurance.

In addition, EPA guidance and state requirements describe actions to take for inactive wells, which EPA refers to as temporarily abandoned wells. These are wells that will not be operating for several months to years. Table 12 shows the different requirements for the eight programs we reviewed.

Table 12. Selected State Requirements for Inactive Wells

State	Description of inactive wells	Frequency of mechanical integrity/pressure tests required for inactive wells
California	An idle well is one that has not been used for fluid injection for a continuous 6-month period during any consecutive 5-year period.	A two-part mechanical integrity test is required for all injection wells. The annulus of each well must be tested every 5 years, and absence of fluid migration must be demonstrated every year.
Colorado	A temporarily abandoned well is one that is incapable of injection without the addition of equipment. Generally, a temporarily abandoned well must be plugged within 6 months, but the state can approve a longer period. A shut-in well is one that is capable of injection by opening valves, activating existing equipment or supplying a power source.	A mechanical integrity test is required within 30 days of temporary abandonment, and within 2 years of shut-in, then once every 5 years for shut-in wells.
Kentucky	A temporarily abandoned well is one that has not been plugged and abandoned that has (1) ceased operation for less than 2 years, or (2) ceased operation for more than 2 years and the operator has provided to EPA a plan to ensure that the well will not endanger underground sources of drinking water.	Requirements for wells apply to temporarily abandoned wells. A mechanical integrity test is required every 5 years.
North Dakota	The removal of injection equipment or the failure to use an injection well for 1 year constitutes abandonment of the well.	Requirements do not distinguish between active and inactive wells. A mechanical integrity test is required at least once every 5 years.

Table 12. (Continued)

State	Description of inactive wells	Frequency of mechanical integrity/pressure tests required for inactive wells
Ohio	None identified	Requirements do not distinguish between active and inactive wells. Certain injection wells must be pressure-tested monthly, or if such monitoring is not feasible, a mechanical integrity test is required every 5 years.
Oklahoma	None identified	Requirements do not distinguish between active and inactive wells. A mechanical integrity is required at least once every five years.
Pennsylvania	A temporarily abandoned well is one that has not been plugged and abandoned that has (1) ceased operation for less than 2 years or (2) ceased operation for more than 2 years and the operator has provided to EPA a plan to ensure that the well will not endanger underground sources of drinking water.	Requirements for wells apply to temporarily abandoned wells. A mechanical integrity test is required every 5 years.
Texas	An inactive well is an unplugged well that has been spudded or has been equipped with cemented casing and that has had no reported disposal, injection, or other permitted activity for a period of greater than 12 months. Plugging operations on inactive wells should commence within a year after drilling or operations have ceased.	Requirements do not distinguish between active and inactive wells. Depending on the type of casing a disposal well has, a mechanical integrity test is required at least once every 5 years and after every workover of the well or as prescribed in the well's permit.

Sources: GAO analysis of federal and state regulations. GAO-14-555

Monitoring and Reporting

To provide an early warning of potential problems, EPA regulations require monitoring of fluids to be injected and well operation. For class II wells, EPA requires, among other things, that operators "monitor the nature of the injected fluids with sufficient frequency to yield data representative of their characteristics;"[94] according to EPA's Basis and Purpose report, such information can help federal and state regulators understand reasons for well failures and take appropriate corrective actions. In addition, class II wells should be monitored on a daily to monthly basis. Table 13 shows the different requirements for the eight programs we reviewed.

Table 13. Select State Requirements for Operator Reporting Requirements for New Wells

State	Injection pressure	Injection volume	Injection fluid	Reporting schedule and monitoring frequency
California	Yes	Yes	Yes	Operators must report monthly on injection pressure, volume, and fluid type. A chemical analysis of injected liquid must be reported whenever its source is changed, or as requested.
Colorado	In some instances	Yes	Yes	Operators must perform a mechanical integrity test. Such test may take the form of an initial pressure test followed by monthly reporting for 60 months of the average casing-tubing annulus pressure. For dedicated injection wells, operators must report monthly on injection volumes. For simultaneous injection wells, operators must report annually on monthly injection volumes. Operators of dedicated injection wells must report monthly the types of chemicals used to treat injection water.

Table 13. (Continued)

State	Injection pressure	Injection volume	Injection fluid	Reporting schedule and monitoring frequency
Kentucky	Yes	Yes	Yes	Operators must report at least annually summarizing requir-ed monitoring, which inclu-des observation of injection pressure and fluid volume weekly for fluid disposal, monthly for enhanced recovery, and daily for liquid hydrocarbon injection; and monthly records of injected fluids and any major changes in their characteristics.
North Dakota	Yes	Yes	Yes	Operators must report monthly.
Ohio	Yes	Yes	None identified	Operators of saltwater injection and enhanced recovery wells must report annually on monthly compilations of averages and maximums based on daily monitoring of injection pressure and volume.
Oklahoma	Yes	Yes [a]	None identified	Operators of noncommercial wells must report annually on monthly records of injection pressure and rate. Operators of commercial wells must file quarterly reports on daily records of injection pressure and monthly records of injection rate.
Pennsylvania	Yes	Yes	Yes	Operators must report at least annually summarizing required monitoring, which includes observation of injection pressure and fluid volume weekly for fluid disposal, monthly for enhanced recovery, and daily for liquid hydrocarbon injection; and monthly records of injected fluids and

State	Injection pressure	Injection volume	Injection fluid	Reporting schedule and monitoring frequency
				any major changes in their characteristics.
Texas	Yes	Yes [a]	None identified	Operators of disposal wells must report annually on monthly records. Operators must report within 24 hours any significant pressure changes indicating leaks in the well.

Sources: GAO analysis of federal and state regulations. | GAO-14-555

[a] More specifically, the injection rate must be reported.

End Notes

[1] EPA regulates six classes of underground injection wells. Class II wells are used to inject brines and other fluids associated with oil and gas production, and hydrocarbons for storage, and are the focus of this report. Additionally, Class I wells are used to inject hazardous wastes, industrial non-hazardous liquids, or municipal wastewater beneath the lowermost underground drinking water sources; Class III wells are used to inject fluids associated with solution mining of minerals beneath the lowermost underground drinking water source; Class IV wells are used to inject hazardous or radioactive wastes into or above underground drinking water sources (these wells are banned unless authorized under a federal or state groundwater remediation project); Class V wells, in general, are used to inject non-hazardous fluids into or above underground drinking water sources, and are typically shallow, on-site disposal systems; Class VI wells are used to inject carbon dioxide for long-term storage.

[2] EPA is also responsible for overseeing, and, in some cases, directly implementing or managing programs in tribal lands and U.S. territories. This report only discusses state programs and EPA-managed programs in states, and does not directly address tribal or territorial programs.

[3] The U.S. Strategic Petroleum Reserve is an emergency stock of oil maintained by the U.S. Department of Energy.

[4] GAO, *Oil and Gas: Information on Shale Resources, Development, and Environmental and Public Health Risks*, GAO-12-732 (Washington, D.C.: Sept. 5, 2012).

[5] In 2005, the Energy Policy Act amended the Safe Drinking Water Act to exempt the underground injection of fluids associated with hydraulic fracturing operations related to oil, gas, or geothermal production activities from regulation under class II programs, except in cases where diesel fuels are used in the fracturing process.

[6] Approximately 18,000 wells were incorporated into the class II program as a result of the reclassification of some wells in California and were not newly drilled wells.

[7] GAO-12-732.

[8] The Energy Information Administration is a statistical agency within the Department of Energy that provides independent data, forecasts, and analyses on energy. Energy Information Administration, *Review of Emerging Resources: U.S. Shale Oil and Shale Gas Plays* (July 2011).

[9] Fossil fuels such as petroleum and natural gas and its derivatives such as plastics, waxes, solvents, and oils, are hydrocarbons.

[10] Enhanced recovery of oil and gas resources can involve many different types of injection, including the injection of steam or gases such as carbon dioxide.

[11] GAO, *Energy-Water Nexus: Information on the Quantity, Quality, and Management of Water Produced during Oil and Gas Production*, GAO-12-156 (Washington, D.C.: Jan. 9, 2012).

[12] Total dissolved solids means the total dissolved (filterable) solids present in a fluid. UIC regulations protect water which has less than 10,000 milligrams per liter (mg/l) total dissolved solids. Most drinking water averages between 200 and 300 mg/l total dissolved solids and water with TDS greater than 500 mg/l is not recommended for human consumption.

[13] Section 1422 of the Safe Drinking Water Act allowed EPA to grant primary enforcement authority to states for all classes of UIC wells.

[14] Section 1425 of the Safe Drinking Water Act created an alternative process for EPA to grant primary enforcement authority to states for class II wells only.

[15] A state must show that its program meets the same four key requirements that the EPA regulations were to address: (1) prohibition of unauthorized injections; (2) authorized and permitted injections must not endanger drinking water sources; (3) include inspection, monitoring, recordkeeping, and reporting requirements; and (4) apply to federal agencies and federal land. 42 U.S.C. § 300h(b)(1) (2014). However, states approved by this process do not need to address each of EPA's technical requirements, such as those related to well construction and testing.

[16] According to EPA, two states with EPA-managed programs, Kentucky and Tennessee, have applications pending to manage their programs.

[17] EPA, *Mid-Course Evaluation of the Class II Underground Injection Control Program: Final Report of the Mid-Course Evaluation Workgroup* (Aug. 22, 1989).

[18] J.B. Smith, U.S. EPA, and L.A. Browning, The Cadmus Group Inc., *Proposed Changes to EPA Class II Well Construction Standards and Area of Review Procedures*. Society of Petroleum Engineers/EPA Exploration and Production Environmental Conference (San Antonio, Texas: March 1993).

[19] Codified at 42 U.S.C. § 300h(d)(1)(B)(ii).

[20] To incorporate state regulations into federal regulations, EPA conducts a rulemaking to codify them into a section of the Code of Federal Regulations. See 40 C.F.R. pt. 147. If EPA decides to enforce state program requirements, EPA must give the state notice; if, after 30 days, the state has failed to commence appropriate action, EPA is to issue an order or begin a court action.

[21] EPA grant funding estimates for state class II programs do not include EPA-managed programs.

[22] GAO, *Funding for 10 States' Programs Supported by Four Environmental Protection Agency Categorical Grants*, GAO-13-504R (Washington, D.C.: May 6, 2013).

[23] We requested staffing data from the states we selected, however, some states do not track staffing data specifically for the class II program. For example, California could only provide estimates of staffing based on the number of class II wells regulated and the total number of oil and gas wells in the state. Similarly, EPA does not track class II program staffing data in Kentucky and Pennsylvania, and could only provide estimates of class II program staffing based on inspection activities in those states. As a result, we did not compare class II program staffing levels across states.

[24] Approximately 18,000 wells were incorporated into the class II program as a result of the reclassification of some wells in California and were not newly drilled wells.

[25] EPA, Office of Drinking Water, *Statement of Basis and Purpose: Underground Injection Control Regulations* (May 1980). This document was intended to summarize the technical basis and purpose underlying the underground injection control regulations promulgated in 40 CFR Part 146, including the safeguards summarized in this report. Our review included a

summary and comparison of the regulations and guidance that establish state and EPA-managed program safeguards, but we did not analyze the technical sufficiency of those safeguards. According to EPA officials, the safeguards established in EPA regulations are still sufficient.

[26] EPA regulations would generally be reflected in state programs approved under the conventional process, in which a state must adopt each EPA requirement or its equivalent. However, all six state programs that we reviewed were approved under the alternative process.

[27] Colorado does not require information on the depth of freshwater or underground sources of drinking water and Texas only requires it for permit applications for disposal wells.

[28] Pennsylvania, Kentucky, and North Dakota also require geologic information on the confining layer above the injection zone. For example, North Dakota has specific language in their class II regulations that requires that the confining layers in the area of review surrounding the zone of injection be free from geologic faults and fractures.

[29] According to North Dakota officials, although state regulations do not require that state officials witness well plugging, in practice, North Dakota officials witness all well pluggings.

[30] For example, when the same operator is responsible for a large number of wells, California, Colorado, Texas, Ohio, and North Dakota allow operators to issue one bond that meets the financial requirements for plugging and abandonment for all of their class II wells.

[31] According to EPA, these funds can be used for both injection and production wells.

[32] 40 C.F.R. §§ 146.23(b), 144.28(g)(2).

[33] In the 1980 *Federal Register* notice announcing its final program regulations, EPA stated that it had evaluated two approaches for directly measuring improvement in groundwater quality from implementation of the UIC program and concluded that neither approach was feasible. The first approach would have drilled groundwater monitoring wells near a random sample of injection wells; this approach was rejected as too expensive and difficult because of the number of monitoring wells that would be needed. The second approach would have monitored groundwater quality using samples from nearby water or other wells; one drawback of this approach, according to EPA, was that it would require groundwater modeling to calculate whether contamination had occurred and EPA determined that such methods and data did not exist. Other drawbacks included the inability to know when contamination started or to distinguish between contamination from an injection well and other potential sources of contamination.

[34] National Research Council, National Academy of Sciences, *Induced Seismicity Potential in Energy Technologies* (Washington, D.C.: 2012).

[35] Groundwater Protection Council, *A White Paper Summarizing a Special Session on Induced Seismicity*, (February 2013).

[36] In early 2013, Ohio started to proactively monitor for induced seismicity prior to and during injection operations near new class II injection wells. Ohio now actively monitors for induced seismicity in real-time at 48 portable seismic stations.

[37] EPA Region 6, *Summary of Area of Review Summit*, (April 2003).

[38] GAO-12-732.

[39] EPA, *Permitting Guidance for Oil and Gas Hydraulic Fracturing Activities Using Diesel Fuels, Underground Injection Control Program Guidance #84* (February 2014), and EPA, *Implementation of the Safe Drinking Water Act's Existing Underground Injection Control Program Requirements for Oil and Gas Hydraulic Fracturing Activities Using Diesel Fuels*, (February 2014).

[40] 40 C.F.R. § 146.24(a)(4)(iii).

[41] Colorado, North Dakota, Ohio, Oklahoma, Pennsylvania and Texas require producers and service companies disclose chemicals used in hydraulic fracturing in FracFocus. In addition, California has developed a Chemical Disclosure Registry that requires disclosure of the

composition, and disposition of hydraulic fracturing fluids within 60 days of the cessation of a hydraulic fracturing operation.

[42] To incorporate changes, EPA conducts a rulemaking to codify state regulations into a section of the Code of Federal Regulations. See 40 C.F.R. pt. 147.

[43] EPA must give the state notice; if after 30 days the state has failed to commence appropriate action, EPA is to issue an order or begin a court action.

[44] The *Federal Register* is the daily publication for rules, proposed rules, and notices of federal agencies.

[45] Simultaneous injection wells separate oil and/or gas from brine inside the wellbore, produces the oil and/or gas along with a small fraction of wastewater to the surface, and reinjects the remaining brine within the wellbore into formations below the base of underground sources of drinking water.

[46] In technical comments on our report, EPA stated that there is only one rulemaking process for both (1) approving state programs and (2) incorporating the state programs into federal regulations. As these examples demonstrate, however, we found many instances in which EPA has approved but not codified state programs.

[47] 40 C.F.R. § 147.1(e).

[48] The Environmental Appeals Board is EPA's final decision maker on administrative appeals under all major environmental statutes that EPA administers.

[49] EPA's Public Water System Supervision program provides grant funding to help eligible states, territories, and tribes develop and implement a Public Water System Supervision program adequate to enforce the requirements of the Safe Drinking Water Act and ensure that water systems comply with drinking water regulations.

[50] Agency officials noted that there are key differences between the class II program and programs such as the Public Water System Supervision program that could raise policy concerns were EPA to consider adopting such an approach in the UIC program. Notably, whereas in the Public Water System Supervision program, EPA regulations apply to public water systems nationwide, under the UIC program, there is no background set of federal regulations that applies nationwide. Rather, there is a different set of regulations in each state, either those adopted by a state and approved by EPA, or in the case of an EPA-managed program, those adopted by EPA.

[51] The Government Performance and Results Act of 1993 requires federal agencies to engage in project management tasks such as setting goals, measuring results, and reporting their progress. To comply with the act agencies are required to prepare annual performance plans that establish the performance goals for the applicable fiscal year, briefly describe how these goals are to be met, and describe how these performance goals can be verified. Agencies must also prepare annual performance reports that review the agency's success or failure in meeting its targeted performance goals. In January of 2011, the act was updated by the Government Performance and Results Act Modernization Act of 2010. This update establishes a new framework that takes a more crosscutting and integrated approach to focusing on results and improving government performance. Additionally, under this new act, agencies are to review and report the results they are achieving more frequently.

[52] The data on class II wells returned to compliance is aggregated with Class III wells, in a way that does not allow class II well data to be separated for reporting purposes.

[53] Some states choose to submit the forms electronically, but the forms are not in a format in which the data can be automatically extracted.

[54] Pub. L. No. 93-523 § 2 (1974), codified as amended at 42 U.S.C. § 300h(a)(1), (b)(1) (2014).

[55] Id., codified as amended at 42 U.S.C. § 300h-1(b)(2) (2014).

[56] 42 U.S.C. § 300h-1(b)(3) (2014).

[57] 42 U.S.C. § 300h-1(b)(1)(B) (2014).

[58] Id.

[59] 42 U.S.C. § 300h-1(b)(2) (2014).

[60] 45 Fed. Reg. 42,500 (June 24, 1980), codified as amended at 40 C.F.R. pt. 146 and other sections; Pub. L. No. 96-502 § 2(a) (1980), codified as amended at 42 U.S.C. § 300h-4 (2014).

[61] 42 U.S.C. § 300h(b)(1) (2014).

[62] H.R. Rep. No. 96-1348 at 5 (Sept. 19, 1980).

[63] Id.

[64] Id. at 6.

[65] 42 U.S.C. § 300h-4(a) (2014).

[66] 42 U.S.C. § 300h-4(c)(2) (2014).

[67] 42 U.S.C. § 300h-4(b) (2014).

[68] Id.

[69] 42 U.S.C. § 300h-1(b)(2) (2014).

[70] 42 U.S.C. § 300h-1(c) (2014).

[71] Id.

[72] 42 U.S.C. § 300h-2(a)(1) (2014).

[73] Id.

[74] 42 U.S.C. §§ 300h-1(d), 300h-1(b)-(c) (2014).

[75] See, e.g., 40 C.F.R § 144.1(f) (2014).

[76] 45 Fed. Reg. 42,500 (June 24, 1980).

[77] Precisely, section 1422 says that states must show that they comply with the added or revised requirement, while section 1425 says that states must demonstrate they meet the statutory requirements as to the changed aspect.

[78] 40 C.F.R. § 145.32(a) (2014).

[79] 40 C.F.R. § 145.32(a) (2014).

[80] The regulation specifies the Administrator, but this has been delegated to the regional administrators.

[81] UIC Guidance #34 at 5.

[82] 48 Fed. Reg. 40,098 (September 2, 1983).

[83] 49 Fed. Reg. 20,138 (May 11, 1984).

[84] 40 C.F.R. §§ 147.1(a), (e) (2014).

[85] That is, section 1423(a)(1) of the act authorizes EPA enforcement of a requirement of an "applicable" UIC program. EPA regulation establishes that part 147 sets forth the applicable UIC programs. Hence, a program not included in part 147 is not an "applicable" program and may not be enforceable by EPA.

[86] 49 Fed. Reg. 20,138 (May 11, 1984). It is unclear why, but part 147 did not codify all state programs that were previously approved. See, e.g., 49 Fed. Reg. at 20,202 (reserving the section for class II), 49 Fed. Reg. 11,179 (March 26, 1984) (approving Arkansas' class II program).

[87] Since the beginning of the program, when EPA has approved state programs, it generally has done so through a rulemaking process or "by rule." Specifically, EPA's process includes a *Federal Register* notice proposing to approve a program, a public comment period, and a *Federal Register* notice announcing the approval. Although EPA established the prerequisite for codification of approved state programs into part 147, EPA generally did not incorporate codification into the approval by rule process.

[88] 61 Fed. Reg. 58,932 (November 16, 1996).

[89] 56 Fed. Reg. 9408 (March 6, 1991).

[90] A shale oil or gas play is a set of discovered or undiscovered oil and/or natural gas accumulations or prospects that exhibit similar geological characteristics.

[91] Energy Information Administration, *Review of Emerging Resources: U.S. Shale Oil and Shale Gas Plays* (July 2011).

[92] EPA, *Mid-Course Evaluation of the Class II Underground Injection Control Program: Final Report of the Mid-Course Evaluation Workgroup.* August 22, 1989 and J.B. Smith, U.S. EPA, and L.A. Browning, The Cadmus Group Inc., *Proposed Changes to EPA Class II Well*

Construction Standards and Area of Review Procedures. Society of Petroleum Engineers/EPA Exploration and Production Environmental Conference (San Antonio, Texas: March 1993).

[93] EPA, Office of Drinking Water, *Statement of Basis and Purpose: Underground Injection Control Regulations* (May 1980).

[94] 40 C.F.R. §§ 146.23(b), 144.28(g)(2).

In: EPA Protection of Drinking Water
Editor: Courtney D. Walsh
ISBN: 978-1-63463-535-6

Chapter 2

DRINKING WATER: CHARACTERIZATION OF INJECTED FLUIDS ASSOCIATED WITH OIL AND GAS PRODUCTION*

U.S. Government Accountability Office

Every day in the United States, at least 2 billion gallons of fluids are injected into underground formations to enhance oil and gas production, or to dispose of fluids brought to the surface during the extraction of oil and gas resources.[1] Wells used for injecting fluids associated with the extraction of oil and gas resources are known as class II injection wells.[2] The majority of fluids injected into class II wells consist largely of brine and may contain pollutants such as chlorides, hydrocarbons, and naturally occurring radioactive materials. The fluids are injected deep underground into porous rock formations, such as sandstone, that are generally below aquifers that can, or do, supply drinking water. Because a significant percentage of the population gets its drinking water from underground aquifers, these wells have raised concerns about the safety of the nation's drinking water.

Domestic production of oil and gas has increased dramatically in the last several years, with corresponding increases in wastewater resulting from production processes.[3] Specifically, hydraulic fracturing combined with horizontal drilling has increased domestic production from unconventional

* This is an edited, reformatted and augmented version of a report, GAO-14-857R, issued September 23, 2014.

sources such as shales, tight sandstones, and coalbed formations. Hydraulic fracturing involves the injection of fluids underground, and fluids that are produced from formations during oil and gas production, including fluids from hydraulic fracturing activities that flow back out of the well, must be disposed of or reused. Oil and gas operations that inject fluids underground for disposal or enhanced recovery are regulated by the Environmental Protection Agency (EPA) under the class II Underground Injection Control (UIC) program. However, the Energy Policy Act of 2005 exempted the process of injecting fluids into a well to hydraulically fracture formations for oil and gas production activities from regulation under the Safe Drinking Water Act.[4]

You requested that we describe what information EPA and states collect from class II well operators on the characteristics of injected fluids. This report describes information that EPA and selected states require injection well operators to provide on the characteristics of fluids injected into class II wells.

To respond to the objective, we reviewed and summarized requirements for operators to provide information about the characteristics of fluids injected into class II wells for the class II program in each of eight selected states. These are outlined in federal regulations, state regulations, guidance, and related program documents for class II programs. To clarify our understanding of the state and federal requirements for class II wells, we interviewed officials in EPA and in each of the selected eight states. We reviewed the same states chosen for our June 27, 2014, review of EPA's class II program: California, Colorado, Kentucky, North Dakota, Ohio, Oklahoma, Pennsylvania, and Texas.[5] These states represent a nongeneralizable sample, selected on the basis of the location of current shale oil- and gas-bearing formations across the country, the number of class II wells in each state, and whether the class II program was managed by the state or EPA regions. To identify current shale oil- and gas-bearing formations across the country, we used Energy Information Administration regions that are organized around national shale oil and shale gas resources.[6] These regions represent diverse geography and geologic formations, as well as different oil and gas and wastewater operations. We selected at least one state in each of the six regions identified by the Energy Information Administration. We also selected states that had higher numbers of class II wells to ensure our sample represented increased class II activity. Finally, we selected states that had both state (California, Colorado, North Dakota, Ohio, Oklahoma, Texas) and EPA-managed programs (Kentucky, Pennsylvania).[7]

We conducted this performance audit from July 2014 to September 2014 in accordance with generally accepted government auditing standards. Those standards require that we plan and perform the audit to obtain sufficient, appropriate evidence to provide a reasonable basis for our findings and conclusions based on our audit objective. We believe that the evidence obtained provides a reasonable basis for our findings and conclusions based on our audit objective.

BACKGROUND

To protect underground sources of drinking water from contamination, class II injection wells are subject to regulation by the UIC program overseen by EPA under the Safe Drinking Water Act.[8] The UIC program regulates three types of class II wells associated with oil and gas production:

1) enhanced recovery wells into which brine, water, steam, carbon dioxide, or other fluids and gases are injected to increase the recovery of oil and gas from oil- or gas-bearing formations;
2) disposal wells into which brines and other fluids brought to the surface during oil and gas production activities are injected for disposal; and (3) storage wells into which liquid petroleum products are injected, generally as part of the U.S. Strategic Petroleum Reserve.[9] The number of class II wells grew from approximately 144,000 in fiscal year 2005 to over 172,000 in fiscal year 2012.[10]

EPA oversees and regulates all classes of injection wells, including class II wells associated with oil and gas production. Under the Safe Drinking Water Act, states may request approval from EPA to manage the UIC programs in their respective borders. Under the act, the EPA Administrator approves states' programs for one or more classes of wells through a rulemaking process, with public notice and comment, and EPA updates federal regulations to reflect the approved program. To gain EPA approval, a state must adopt and implement a program that meets the minimum requirements established in EPA regulations,[11] or, specifically for class II programs, a state may adopt and implement a program that it demonstrates to be as effective as federal regulations in preventing the contamination of underground sources of drinking water.[12] Once EPA has approved a state's program, the state has primary management and enforcement responsibility for its UIC program, known as primacy. In states that do not have approval to manage their programs, EPA regional offices manage the programs in the states directly.[13]

EPA has approved 39 states, either through the conventional or alternative process, to manage their own class II programs, while EPA manages the class II programs in 11 states.

EPA developed safeguards to protect underground drinking water sources in the 1980s with the purpose of preventing fluids that are injected into underground formations from endangering underground drinking water sources.[14] These safeguards require well operators to, among other things, meet technical standards for constructing, operating, testing, and monitoring injection wells. As part of these safeguards, EPA regulations require that operators "monitor the nature of the injected fluids with sufficient frequency to yield data representative of their characteristics."[15] According to EPA officials, the purpose of monitoring is to ensure that the fluids to be injected into a class II well are fluids that are allowed to be injected into a class II well and are considered nonhazardous. According to EPA documentation, information on the characteristics of injected fluids can help to provide an early warning of potential contamination of underground sources of drinking water and can help federal and state regulators understand the reasons for well failures and take appropriate corrective actions.[16]

States, and EPA regional offices that manage programs in some states, are largely responsible for the day-to-day management of the class II program. Management includes permitting wells, inspecting wells, enforcing regulations, developing and applying guidance, and collecting and reporting program data to EPA.

Information Collected by EPA and Selected States on the Characteristics of Fluids Injected into Class II Wells Varies

Class II programs in seven of the eight states we selected require that permit applicants provide some information on the characteristics of fluids injected into class II wells prior to permitting, but the specificity and frequency of the information applicants report varies from state to state. Specifically, all of the states we selected except for Ohio require that applicants provide some information on the characteristics of fluids injected into class II wells, but the specific constituents to be reported differ by state. For example, EPA Region 4, which manages the class II program in Kentucky, requests that applicants analyze the pH, total dissolved solids, and specific gravity of the fluid to be injected into a class II well, and can request that applicants provide additional data on the chemical characteristics of injected fluids. Conversely, class II programs in Colorado and North Dakota request an analysis of the fluids injected into class II wells but do not specify what information on fluid characteristics applicants are required to provide. While Ohio's regulations do not require operators to provide information on the characteristics of fluids injected, the regulations narrowly define what fluids can be injected into class II wells. In addition, according to state officials, Ohio conducted research on the characteristics of produced water in the state's oil and gas producing formations and samples fluids injected into class II wells during well inspections.

According to EPA officials, fluid characterization requirements for class II wells are designed to ensure that no chemicals are injected that could potentially damage the wells. In addition, EPA officials told us that the agency does not prescribe a set list of constituents that state and EPA-managed class II programs should monitor. Specifically, according to guidance issued by EPA on class II well monitoring, EPA cannot recommend a single set of constituents that should be monitored because injection fluids can vary widely in composition and contain different naturally occurring chemicals and fluids used in oil and gas production depending on the source of the injection fluid. As a result, state programs and programs managed by EPA regions have discretion to monitor the injection fluid constituents that they deem critical to protect underground sources of drinking water in their respective states or regions.

In addition, five of the eight programs we selected request that well operators conduct additional analyses of fluids injected into class II wells after the well has been permitted. Of these five programs, in Kentucky, EPA Region 4 requests that operators provide updated information on fluids injected into class II wells annually, and Colorado requests updated reporting every 5 years. Texas and California request updated information on fluids injected when the source of the injected fluids changes. EPA Region 3, which manages the program in Pennsylvania, requests that operators provide updated information on fluids injected into commercial class II wells every year and fluids injected into private wells every 2 years. Of the remaining states, North Dakota and Oklahoma only request information on fluid characteristics during the class II well permitting process or when requested by state officials. Ohio regulations allow state officials to sample injection fluids at any time during a well's operation.

Table 1 presents detailed information requested from operators by select states and EPA regional offices on the characteristics of fluids injected into class II wells.

Table 1. Information Collected by EPA and Selected States on the Characteristics of Fluids Injected into Class II Wells

State, Agency	Fluid characteristics reported	Frequency of testing (or reporting)
California, Division of Oil, Gas, and Geothermal Resources	According to officials, the chemical analysis typically lists the major constituents of the injection fluid, total dissolved solids, and pH.	The source and analysis of the injection fluid is required in any injection project application.[a] After the project is approved, a chemical analysis of the liquid being injected should be provided by the operator whenever the source of injection fluid is changed, or when the division requests the data from the operator.[b]
Colorado, Oil and Gas Conservation Commission	According to officials, injected fluids must be analyzed for total dissolved solids, major cations and major anions.[c] Officials stated that they include these conditions in permit approval letters.	According to officials, water analyses of fluids injected into injection wells are required within 1 year of commencement of injection. Injected fluids are required to be monitored with sufficient frequency to yield data representative of their characteristics. Therefore, water analyses of fluids injected are required as a Condition of Approval, on Form 26 (Source of Produced Water For Disposal), when sources are added or deleted. Water analyses of fluids injected are also required at 5-year intervals.
Kentucky, EPA Region 4	According to officials, the fluid analysis must include the pH, total dissolved solids, and specific gravity; as well as a list of all chemicals and their composition used for any well stimulation and fracturing during that sampling year, and a list of any additives	According to officials, analyses shall be made beginning within 12 months from the effective date of the permit, or 12 months from the most recent analysis, whichever is later. For wells that resume injection

State, Agency	Fluid characteristics reported	Frequency of testing (or reporting)
	used and their chemical composition, including any inhibitors used to prevent scaling, corrosion, or bacterial growth, and these lists should indicate the brand name of the product and the manufacturer. On EPA's written request, an injection fluid analysis can include the following additional constituents: barium, calcium, total iron, magnesium, sodium, bicarbonate, carbonate, chloride, sulfate, carbon dioxide, dissolved oxygen, hydrogen sulfide, and purgeable aromatic hydrocarbons.	after having been shut-in, the permittee will have 30 days from the date injection resumes for the submission of the injection fluid analysis.[d]
North Dakota, Industrial Commission, Oil and Gas Division	Only fluids brought to the surface in connection with natural gas storage, or conventional oil or natural gas production, and wastewater from gas plants that are integral to production operations may be injected into a class II well.[e] In the permit application, a quantitative analysis from a state-certified laboratory of a representative sample of the fluid to be injected is required.[f] This analysis reports the major anions and cations.	After the permit is approved, the operator must conduct sampling as the commission may require.[g]
Ohio, Department of Natural Resources, Division of Oil and Gas Resource Management	According to officials, class II injection well operators are not required to analyze the chemical composition of injected fluids. Only saline geological formation water resulting from, obtained from, or produced in connection with exploration, drilling, stimulation, production, or plugging of oil and gas wells (defined in Ohio law as "brine"[h]) is permitted to be injected into class II saltwater injection wells.[i] According to officials, the division	The division may sample injection fluids at any time during injection operations.[k] According to officials, division inspectors collect fluid samples from permitted injection facilities for analysis at an EPA certified laboratory.

Table 1. (Continued)

State, Agency	Fluid characteristics reported	Frequency of testing (or reporting)
	has supported research to analyze produced water samples from oil and gas producing formations.[j] In addition, operators may monitor the specific gravity of fluids when applying to increase permitted injection pressure limits.	
Oklahoma, Corporation Commission	In the permit application, operators will provide an analysis of sodium, chloride, and total dissolved solids.[l]	According to state officials, fluids may be tested during pollution investigations or as a result of citizen complaints.
Pennsylvania, EPA Region 3	According to EPA officials, in the permit application, operators are required to do a fluid analysis including: specific gravity, total organic carbon, pH, specific conductance, sodium, chloride, iron, manganese, total dissolved solids, barium, hydrogen sulfide, alkalinity and dissolved oxygen, hardness, and magnesium to characterize the fluids to be injected.	According to EPA officials, operators are required to provide an analysis of injected fluids during permitting. Commercial wells are required to resample every year and whenever the operator anticipates a change in the source of the fluid to be injected. Privately owned wells should be resampled every 2 years and whenever the source changes.
Texas, Railroad Commission	According to officials, during permitting, an operator provides general categories of fluid types, such as exempt oil and gas waste, fresh water, salt water, fracture water flowback, natural gas, hydrogen sulfide, steam, air polymer, carbon dioxide, nitrogen, and naturally occurring radioactive material as defined by Texas law.	According to officials, a permit amendment is required if the operator wants authority to inject a fluid that is different from that approved in the permit.

Sources: GAO analysis of EPA and state regulations and guidance. | GAO-14-857R

[a] Cal. Code Regs. tit. 14 § 1724.7(c)(7) (2014).

[b] Cal. Code Regs. tit. 14, § 1724.10(d) (2014).

[c] A cation is a positively charged atom or molecule. An anion is a negatively charged atom or molecule.

[d] Shut-in wells are wells that are shut off for a period of time and do not accept injected fluids.
[e] See N.D. Admin. Code 43-02-05-01(2)(a) (2014).
[f] N.D. Admin. Code 43-02-05-04(1)(m) (2014).
[g] N.D. Admin. Code 43-02-05-12(4) (2014).
[h] Ohio Rev. Code Ann. § 1509.01(U) (2014).
[i] According to Ohio officials, samples were analyzed for total dissolved solids, specific conductance, pH, specific gravity, sodium, potassium, calcium, magnesium, manganese, iron, chloride, sulfate, alkalinity, strontium, lithium, aluminum, cadmium, chromium, cobalt, copper, lead, mercury, nickel, titanium, and zinc.
[j] See Ohio Admin. Code 1501:9-3-06 (2014).
[k] Ohio Admin. Code 1501:9-3-07(J) (2014).
[l] Okla. Admin. Code § 165:10-5-5(b)(5)(D) (2014).

AGENCY COMMENTS

We provided EPA and the managers of the eight state class II programs we reviewed with a draft of this report for comment. EPA and Ohio provided technical comments that were incorporated, as appropriate.

J. Alfredo Gómez
Director,
Natural Resources and Environment

End Notes

[1] Environmental Protection Agency (EPA), Class II Wells – Oil and Gas Related Injection Wells, accessed September 08, 2014, http://water.epa.gov/type/groundwater/uic/class2/index.cfm.
[2] EPA regulates six classes of underground injection wells. Class II wells are used to inject brines and other fluids associated with oil and gas production, as well as hydrocarbons for storage, and are the focus of this report. Additionally, class I wells are used to inject hazardous wastes, industrial non-hazardous liquids, or municipal wastewater beneath the lowermost underground drinking water sources; class III wells are used to inject fluids associated with solution mining of minerals beneath the lowermost underground drinking water source; class IV wells are used to inject hazardous or radioactive wastes into or above underground drinking water sources (these wells are banned unless authorized under a federal or state groundwater remediation project); class V wells, in general, are used to inject nonhazardous fluids into or above underground drinking water sources, and are typically shallow, on-site disposal systems; class VI wells are used to inject carbon dioxide for long-term storage.
[3] GAO, Oil and Gas: Information on Shale Resources, Development, and Environmental and Public Health Risks, GAO-12-732 (Washington, D.C.: Sept. 5, 2012).

[4] In 2005, the Energy Policy Act amended the Safe Drinking Water Act to exempt the underground injection of fluids associated with hydraulic fracturing operations related to oil, gas, or geothermal production activities from regulation under class II programs, except in cases where diesel fuels are used in the fracturing process.

[5] For additional information on the state selection methodology, see GAO, Drinking Water: EPA Program to Protect Underground Sources from Injection of Fluids Associated With Oil and Gas Production Needs Improvement, GAO-14-555 (Washington, D.C.: June 27, 2014).

[6] Energy Information Administration, Review of Emerging Resources: U.S. Shale Oil and Shale Gas Plays (July 2011).

[7] States request approval from EPA to manage the UIC programs within their respective borders, including the class II programs. Under the Safe Drinking Water Act, the EPA Administrator approves state programs for one or more classes of wells through a rulemaking process, with public notice and comment, and EPA updates federal regulations to reflect the approved program. Once EPA has approved a state's program, the state has primary management and enforcement responsibility for its UIC program, known as primacy. In states that do not have approval to manage their programs, EPA regional offices manage the programs in the state directly. Twenty-five states with class II wells manage their class II programs and regulate over 95 percent of the class II wells nationwide. Eight states with class II wells have programs managed by EPA regional offices. The remaining 17 states have no class II wells.

[8] An underground source of drinking water as defined in 40 C.F.R. § 144.3 is any nonexempt aquifer or part of such an aquifer which: supplies any public water system; or which contains a sufficient quantity of groundwater to supply a public water system, and (1) currently supplies drinking water for human consumption; or (2) contains fewer than 10,000 milligrams/liter of total dissolved solids.

[9] The U.S. Strategic Petroleum Reserve is an emergency stock of oil maintained by the U.S. Department of Energy.

[10] Approximately 18,000 wells were not newly drilled wells, but were incorporated into the class II program as a result of the reclassification of some wells in California.

[11] Section 1422 of the Safe Drinking Water Act authorizes EPA to grant primary enforcement authority to states for all classes of UIC wells.

[12] Section 1425 of the Safe Drinking Water Act created an alternative process for EPA to grant primary enforcement authority to states for class II wells only.

[13] For additional information on class II wells, see GAO-14-555. Some states that have been approved for primacy do not have class II wells, and the number of states with class II programs is smaller than the number with primacy approved.

[14] EPA, Office of Drinking Water, Statement of Basis and Purpose: Underground Injection Control Regulations, (Washington, D.C.: May 1980). In GAO-14-555, our review included a summary and comparison of the regulations and guidance that establish state and EPA-managed program safeguards, but we did not analyze the technical sufficiency of those safeguards. According to EPA officials, the safeguards established in EPA regulations are still sufficient.

[15] 40 C.F.R. §§ 146.23(b), 144.28(g)(2).

[16] EPA, Office of Drinking Water, Statement of Basis and Purpose: Underground Injection Control Regulations, (Washington, D.C.: May 1980).

In: EPA Protection of Drinking Water
Editor: Courtney D. Walsh
ISBN: 978-1-63463-535-6

Chapter 3

STATEMENT OF BASIS AND PURPOSE: UNDERGROUND INJECTION CONTROL REGULATIONS*

Environmental Protection Agency

STATEMENT OF BASIS AND PURPOSE INTRODUCTION

This document is intended to summarize the basis and purpose underlying the underground injection control regulations promulgated in 40 CFR Part 146. It sets forth generally the reasoning behind the Agency's regulatory choices and references data upon which EPA relied.

This statement first discusses the categorization of wells. It then surveys the major pathways which contaminants can take to enter underground sources of drinking water and the requirements which these regulations impose to assure that movement of fluids resulting from well injection does not contaminate underground sources of drinking water. Programmatic requirements of Part 146, such as monitoring and reporting, are covered in the concluding section.

In general, the requirements of the regulation differ from those of the initial proposal of this regulation (*41 FR 36730*, August 31, 1976) and the reproposal of this regulation (*44 FR 36730)* in that they furnish a greater

* This is an edited, reformatted and augmented version of a document (docket control number D 01079), issued by the EPA Office of Drinking Water, May 1980.

degree of flexibility to State Directors in regulating well injection. EPA has modified its earlier proposal in this way as it became more fully aware of various well injection practices, the characteristics of substrata into which fluids are injected, and the range of methods by which well injection is accomplished.

CATEGORIZATION OF WELLS

The regulations separate wells into distinct categories. This categorization is necessary to assure that wells with common design and operating techniques will be required to meet appropriate performance criteria.

In categorizing wells, EPA first looked to available literature regarding the injection practices. It considered information on existing and abandoned injection well practices, well construction technology, and on the variety of fluids injected into wells. It discussed with regulatory agency personnel from many States their experience with then-existing well injection regulatory practices and reviewed existing regulations in a number of States.[1] After such review and discussions, it commissioned a reputable consulting organization to provide assessments and a report on types of wells and their typical operation.[2] EPA then studied this information to arrive at a consistent and comprehensive well classification scheme. As a result, EPA decided to classify wells into the following five groups:

- Class I: Wells used by generators of hazardous wastes or owners or operators of hazardous waste management facilities to inject hazardous waste, other than Class IV wells. other industrial and municipal disposal wells which inject fluids beneath the lowermost formation containing, within one quarter mile of the well bore, an underground source of drinking water.
- Class II: Wells which inject fluids (1) which are brought to the surface in connection with conventional oil or natural gas production (2) which are used for enhanced recovery of oil or natural gas and (3) which are used for storage of hydrocarbons which are liquid at standard temperature and pressure.
- Class III: Wells which inject for extraction of minerals or energy, including: mining of sulfur by the Frasch process; solution mining of

minerals; in-situ combustion of fossil fuel, and recovery of geothermal energy.

- Class IV: Wells used by generators of hazardous wastes or of radioactive wastes, by owners or operators of hazardous wasce management facilities or by owners or operators of radioactive wastes disposal sites to dispose of hazardous wastes into or above a formation which within one quarter mile of the well contains an underground source of drinking water.
- Class V: Injection wells not included in Classes I, II, III, or IV.

In formulating these classifications, EPA gave substantial weight to a number of considerations. First the Agency concluded that wells which inject into strata nearest thd land surface should, as a general matter, be classified separately from those which inject into strata at greater depth. The method of injection which wells use is frequently dependent upon the injection horizon into which they deposit fluids. Wells which inject into strata near the land surface often inject by use of simple gravity. often crudely constructed, they can simply be holes dug or bored into the ground, the sides of which may be stabilized by brick, stone, timber, or other materials in the well. They can function as convenient dumping sites for wastes, or, in other instances, can serve beneficial purposes, such as recharging groundwater supplies or creating a subsurface barrier to saltwater intrusion.[3]

Wells which inject into lower strata are usually constructed and operated differently from wells which inject into strata near the land surface. Such wells are drilled rather than dug or bored, and emplace fluids into the subsurface by use of more sophisticated technology, materials, and equipment. Wells of this sort require the use of casing and cementing.[4] Escape of injected fluids into sources of drinking water is prevented by such casing, and by tubing and packer or other methods. Injection is accomplished by either the force of gravity or the application of additional mechanical pressure to overcome the natural friction and hydrostatic, resistance of the receiving formation.[5]

In addition, aquifers nearest the land surface most often supply water for domestic use.[6] Consequently, wells which inject into or above these aquifers increase the risk of human exposure to the injected contaminants.

These considerations influenced the categorization of wells in Classes I and II separately from those in Classes IV and V. Classes I and II encompass wells which normally inject into strata below underground sources of drinking water. Class I wells are further limited to those which inject beneath formations which contain an USDW within 1/4 mile of the well site; other wells are included in Class V. The agency chose this approach since individual formations may be identifiable for hundreds of miles and a formation which is suitable in one area as a source of drinking water may not be in other sections. This limitation prevents a well from being subjected to Class I requirements simply because it injects under an aquifer which, miles away, contains drinking water. Class IV wells (which by definition inject into or above strata containing underground sources of drinking water) will generally inject into or above the aquifers nearest the land surface. Class V for the most part comprises wells which inject non-hazardous materials into those same aquifers.[7]

Also influencing this proposed well class-ification was the nature of injected fluids. Wells which handle hazardous materials war-rant close regulatory scrutiny. This consideration influenced EPA to create a separate category (Class I) for wells which dispose of industrial and municipal wastes. Such wastes commonly contain chemicals or other substances which can be fairly characterized as noxious and, as appropriate, require separate performance criteria. Nuclear waste practices are currently being reviewed by the Administration and therefore EPA is not including wells which inject nuclear waste a USDW into a Class I at this time.[8] Until a final policy is developed, such wells are included in Class V.

Similarly, among those wells which inject into aquifers nearest the land surface. Class IV wells were separated from Class V because of the heightened risk which Class IV wells create. Class IV wells may be the most harmful class of wells because of the hazardous nature of fluids injected into them and the proximity of their injection zone to underground sources of drinking water. Within wells categorized as Class IV, two general subcategories may be defined: those which inject hazardous waste directly into underground sources of drinking water, and those which inject hazardous waste above underground sources of drinking water. It was felt that this difference was sufficient to warrant-distinct treatment. Accordingly, Class IV wells which inject into an underground source of drinking water are being banned under an "interim final" rule within 6 months after a State UIC program becomes effective. A decision regarding Class IV wells which inject above an underground source of drinking water is being reserved-at this time pending further Agency consideration and public comment. The rational for this approach is presented in more detail in a later discussion. Management of radioactive waste which is injected into or above USDWs will follow this same regulatory scheme, i.e., such waste injected into a underground source of drinking water will be banned within 6 months after a State UIC program becomes effective, and decisions regarding such waste injected above USDWs are being reserved. This applies to radioactive waste as defined by both RCRA and the Atomic Energy Act of 1954.

An additional factor which influenced this reproposed classification of wells was the Safe Drinking Water Act ("SDWA") itself. Sections 1421(b),(2) and 1422(c) of the SDWA state that regulations for State underground-injection control programs may not prescribe requirements which interfere with or impede underground injection in connection with oil and natural gas production or the secondary or tertiary recovery of oil and gas production unless such requirements are essential to assure that underground sources of drinking water will not be endangered by such injection. House Report No. 93-1185 accompanying the Act takes care to clarify this directive. At page 31, the Report characterizes the term "interferes with or impedes" as referring to only those requirements which could "stop or substantially delay" oil or natural gas production. Thus, the "test" of essentiality would only be relevant upon a demonstration that a requirement would stop or substantially delay such production.

EPA has observed this statutory admonition by including all injection wells relating to oil and natural gas production and *hydrocarbon* storage into a single category (Class II) with the exception of gas storage wells. Gas storage wells have been included in Class V wells. Such a grouping makes it possible to modify specific requirements and allow additional flexibility where possible without endangering human health. It was felt that the economic incentive for preventing leakage, and the relatively innocuous character (natural gas is not highly soluble) associated with gas storage wells warranted that these wells be studied further before subjecting them to regulation equivalent to other Class II wells.

Class III, which includes special process wells (including those used for solution mining) are classified separately from other wells because of their atypical injection practices. Special process wells serve a variety of purposes, including the extraction of minerals or other materials from the earth. Individual domestic injection wells used to generate heat and/or electricity, although special process wells, have been included in Class V. The Agency chose to consider these systems under Class V due to the limited impact on underground sources of drinking water anticipated from these wells. It should be noted that domestic syst ems utilizing a closed loop in conjunction with heat exchangers are not injection wells and consequently, are not covered under these regulations.

In the course of public review, a number of commenters urged that Class III wells be subcategorized. In response, the Agency commissioned a study to provide information on this issue.[9] Based on this study it does not appear that either current operating and construction practice for existing State regulations governing such wells differentiate sufficiently to warrant subcategorizing Class III wells. Consequently, the final regulations do not establish subcategories for Class III wells.

As stated earlier, Class IV wells injecting into a USDW are banned; decisions concerning Class IV wells injecting above an USDW are being reserved. In order to afford some level of protection before the final promulgation of regultions for Class IV wells and implementation of State UIC programs, all injectors of hazardous waste must obtain "interim Status" under the RCRA hazardous waste management program. The Agency decided to reserve final decisions concerning Class IV wells which inject above USDW's for several reasons. Commenters pointed out that injection wells which overlie deep or remote drinking water supplies would have little or no potential for contaminating the aquifer and thereby endangering health. Moreover, portions of some aquifers may be so deep or so remote that they may never serve as drinking water sources, or may not be subject to contamination from injection practices.

A further reason for the proposed approach is that regulations under RCRA and SDWA overlap at several points. Facilities under Class I and Class IV overlap the class of facilities designated under RCRA as hazardous waste management facilities. It is, therefore, appropriate that technical standards under RCRA and UIC be consistent, to the extent allowable under the governing statutes, for facilities capable of causing a similar degree of risk.

EPA anticipates issuance of permitting standards for HWM facilities until fall 1980. Adoption of UIC standards now for class IV wells could prove misleading to the States and the public, because EPA might decide this fall to revise the standards to reflect policy decisions made in connection with RCRA standards. Accordingly, EPA has determined that the best course is to defer the "technical standards for Class IV wells which inject above USDWs until fall 1980.

Finally, Class V wells include all wells not covered by the preceding categories, and those wells for which EPA currently lacks comprehensive information. With the exception of wells injecting radioactive waste, which are covered under Class IV and V, the Agency has reason to believe that Class V wells pose a significantly lesser environmental danger than do other categories of wells. Some Class V wells can cause risks to public health, of course, but many of them can be actually beneficial to groundwater. Due to incomplete data, EPA has classified these remaining wells together and is proposing no immediate performance criteria for them at this time. Instead, these wells are to be assessed and, based on that assessment, EPA will formulate a regulatory program suitableior them at a later time. In the meantime, if remedial action appears necessary, an individual permit may be required (§122.37) or the Director may require remedial action or closure by order (§122.37(c)).

PERFORMANCE CRITERIA

The regulations propose the use of a variety of measures to assure that injection wells will not jeopardize underground sources of drinking water. This section addresses, the major technical requirements by discussing each in conjunction with the particular problem it is designed to prevent or remedy. The "problems" are basically six in number, and are described here as "pathways of contamination" - ways in which fluids can escape the well or injection horizon and enter underground sources of drinking water. These "pathways" are the following:

1) movement of fluids through a faulty injection well casing;
2) movement of fluids through the annulus located between the casing and well bore;
3) movement of fluids from'an injection zone through the confining strata;
4) vertical movement of fluids through improperly abandoned and improperly completed wells;
5) lateral movement of fluids from within an injection zone into a protected portion of that stratum; and
6) direct injection of fluids into or above an underground source of drinking water.

PATHWAY 1 - MIGRATION OF FLUIDS THROUGH A FAULTY INJECTION WELL CASING

The casing of a well can serve a variety of purposes. It supports the well bore to prevent collapse of the hole and consequent loss of the well, serves as the conductor of injected fluids from the land surface to the intended injection zone, and supports other components of the well. If a well casing is defective, injected fluids may leak through it. Such migration can contaminate an underground source of drinking water.[10]

To prevent migration of fluids in this manner, the promulgated regulations require that wells in Class I use casing sufficient to prevent the movement of fluids into any underground source of drinking water. Casing requirements for Class II and III wells are different and are discussed in more detail below.[11] The impact of this standard should vary on a well-by-well basis. In some instances, injection wells would only need a minimal surface casing to prevent migration of fluids into underground sources of drinking water. In other cases, multiple strings of casing might be necessary. EPA is proposing this flexible, goal-related standard, rather than a fixed requirement, in order to allow State Directors the discretion to vary the requirement, as appropriate, in each instance. Allowing this discretion should lessen the cost of the requirement while still accomplishing its preventive objective.[12]

The regulations also require wells to comply with certain operational requirements which can serve to minimize migration of fluids through casing. Foremost among these are the requirements to demonstrate mechanical integrity.[13] A mechanical integrity test is used to verify, as its name indicates, the "integrity" of a well, i.e., whether there is an absence of significant leaks.[14]

The determination of what constitutes a significant leak is left to the Director. This acknowledges the site-specific nature of the question and allows a case-by-case review of important local phenomena that must be considered in establishing-"significance". The regulations require operators of all new Class I-III wells (wells coming into operation after an applicable UIC program becomes effective in the State) to conduct mechanical inte§rity tests and provide the results to the Director. If a test indicated that a well did not have mechanical integrity, i.e., it leaked injected fluids, the well would not be authorized for injection. For existing-wells, the regulations require that mechanical integrity be demonstrated before continued operation of the well can be authorized.

The selection of a mechanical integrity test as a requirement of these regulations is uniquely appropriate because normally wells cannot be inspected directly. Well integrity can be demonstrated by the absence of a significant leak in the casing, tubing or packer and the absence of significant fluid movement into an underground source of drinking water. The regulations specify a choice of two tests to detect leaks, and two others to detect fluid movement.[15] The regulations specify monitoring of annulus pressure or pressure tests'with a liquid or gas for detecting leaks, and use of noise or temperature logs for detecting fluid movement. Existing Class II wells may use well records as proof of adequate cement to prevent fluid movement. These tests are commonplace in the well injection, industry, and are considered reliable indicators of mechanical integrity.[16]

The regulations also would allow the-use of mechanical text.[17] To integrity tests not specifically listed in the t use any of these tests, a Director would have to demonstrate its suitability for the intended. purpose and secure EPA approval prior to its use. Once approved by EPA, the test would be eligible for use by all persons unless specifically restricted. EPA allows this flexibility because it recognizes that there may be mechanical integrity tests which, although unspecified in the regulations, are fully adequate to detect well defects. Moreover, tests which might be acceptable may be developed in the future.

The regulations further require that operators of wells which have been authorized for injection under this program perform, additional, mechanical integrity testing at least once every five years of operation for most wells.[18] However, additional mechanical integrity tests for Class III wells will only be required for those wells which are used for relatively long periods, such-as salt solution and geothermal wells. Other Class III wells, which have a shorter life span, will not be required to perform periodic mechanical integrity tests. In addition, Class II wells may use well records to demonstrate the presence of adequate cement to prevent significant fluid movement. As part of an evaluation, a statistically valid random sample of wells will be tested with either noise or temperature logs to assess the adequacy of well records as a measurement of mechanical integrity.

The Agency decided on the five year frequency period after long consideration and consultation with state officials. EPA staff determined that the requirement for a mechanical integrity test at least every five years during operation of the well would provide satisfactory assurance of continued well soundness and in addition would be reasonable from a cost perspective. Moreover, the five-year review schedule facilitates Agency efforts to combine the several permit programs under its charge.

A second protective feature of these regulations is the requirement for a tubing and packer, fluid seal, or an alternative approved by the director for Class I wells. The Agency applied this requirement to Class I wells due to the potentially corrosive nature associated with Class I wastes. This requirement does not apply to municipal. wells injecting non-corrosive fluids. The reproposal specified only tubing and packer or alternative. As a result of public comment, the Agency included the use of a fluid seal in the final regulations. Fluid seals are used extensively and have proved effective. Tubing and packer can best be described as a removable liner device within a well which isolates the casing of the well from injected fluids. By preventing this contact between casing and injected fluids, the possibility of movement of contaminants through leaks in the casing is greatly diminished. For the same reason, tubing and packer or equivalent also lessens the chances of corrosion of the casing. Tubing and packer offers two further advantages. It isolates the annulus (between the tubing and casing) from the injection zone, facilitating detection of any leaks in the tubing. It also allows for visual inspection for deterioration of the tubing during routine maintenance.[19]

The regulations make the use of tubing and packer or an acceptable substitute mandatory for Class I well S,[20] except for municipal wells injecting only non-corrosive wastes. EPA expects that Class I wells will be injecting highly corrosive material more frequently than Class II or III wells,[21] hence, routine use of tubing and packer or an acceptable substitute becomes appropriate (For Class II and III wells, the requirement to use tubing and packer is discretionary with the Director because the inflexible use of the requirement for Class II and III wells would likely interfere with production from many of these wells without any significant benefit to protecting USDW).[22] Even though a tubing and packer requirement is not mandatory for wells in Classes II and III, Directors should require its use when appropriate to prevent fluid migration into underground sources of drinking water.)

When the use of a packer in Class I wells is inappropriate, the regulations allow for use of alternative means to accomplish the'saime objective provided that the Director approves such methods.[23] In fact, based upon the type of well involved, it is possible that an alternative to tubing and packer or fluid seal might actually provide a greater degree of protection.[24] When other effective methods are proposed, EPA does not oppose their use. Prior to use, however, EPA reserves the right of review and approval.

The final provision by which the regulations propose to eliminate contamination through this first pathway is to require that Class I and Class III wells-which in ect corrosive fluids be constructed of corrosion'-resistant materials. [25] This standard is intended to prolong the operating life and continued viability of wells.

Pathway 2 - Migration of Fluids Through the Annulus Located Between the Casing and the Well Bore

A second way by which contaminants can reach underground sources of drinking water is by migrating upward through the annulus located between the drilled hole and the casing. Under usual injection conditions, injected fluids, upon leaving the well in the injection zone, enter a stratum which to some degree resists the entry of the fluids. Resistance results from friction created by extremely small openings in the materials which comprise the injection zone.

Because fluids tend to take the path of least resistance unless properly contained, they may travel upward through this annulus. If sufficient injection pressure exists, the fluids could migrate into an overlying source of drinking water.

The measures taken in the regulations to prevent contamination by this pathway are parallel to those already mentioned concerning Pathway 1. In this case, well injectors must demonstrate to the satisfaction of the Director that there is no significant fluid movement into underground sources of drinking water through this annulus. Mechanical integrity tests can be conducted to provide information on contamination by this route.[26] As with Pathway 1, and for the same reasons, mechanical integrity must be demonstrated at least every five years.

For Class I and III wells, the annulus between the hole and casing must also be filled with cement adequate to prevent the flow of fluids into an overlying drinking water source.[27] Depending upon the complexity of the well, this cementing can be accomplished in different ways. A well with a single casing, for example, may need cementing at only one interval (e.g., through the confining stratum which separates the injection zone from the source of drinking water). Other wells, which penetrate to greater depths or which involve more than one casing, may need a more elaborate cementing procedure. Because of this range, EPA is proposing the cementing requirement in general terms and intends to leave decision making to Directors' discretion. Directors are instructed in the regulations to take a variety-of factors into account when determining specific cementing requirements for individual wells.[28]

All new Class II wells will be subject to requirements outlined above. Existing and converted Class II wells need nut meet these requirements if they were subject to regulatory controls at the time they were drilled and they are in compliance with those controls, and injection will not result in the migration of fluids into an underground source of drinking water so as to create a significant risk to the health of persons using the source as drinking water. Similarly, new (newly drilled) wells in existing fields must meet casing and cementing requirements applicable to the field, and cannot allow movement of fluids into an underground source of drinking water if such movement will create a significant risk to health of persons.[29]

For Class III wells, all new wells must comply with the requirements discussed above. Existing wells which have long lives, such as salt solution and geothermal wells, must demonstrate mechanical integrity; however, they are not required to meet other casing and cementing requirements. Various considerations underlie this modified approach.

As mentioned in the preamble to these regulations, costs played a role: EPA data indicates that compliance for Class II wells equivalent to casing and cementing requirements for Class I wells could generate cos ts to the oil industry of more than $20 billion over 5 years.30, [31] Imposing regulatory requirements of this financial magnitude in EPA's view, would interfere with injection of brines or other fluids which are brought to the surface in connection with Oil and natural gas production and with injection for secondary or tertiary recovery of oil or natural gas without being essential to assure that underground sources of drinking water will not be endangered by such injection. Moreover, the imposition of this casing and cementing requirement could be an unnecessary disruption of state UIC programs currently in effect and being enforced in a substantial number of states.

In addition, the imposition of the "full" casing and cementing requirement on Class II wells in existing injection fields would not yield significant environmental benefit. If past injection was performed in an unsafe way, nearby water resources will likely be too contaminated for consumption as drinking water. Imposing casing and cementing in this instance would not be helpful to the environment. On the other hand, if the injection has been performed historically in a way which is protective of underground drinking water, it is reasonable to believe that the injection method will continue to protect underground sources of drinking water. These facts are particularly applicable to Class II wells because they are relatively older than wells in other categories[32] and are normally found in groups the members of which are similarly constructed.[33] Older wells, with longer histories of operation; are more likely to have contaminated drinking water, if at all, by this time, than are newer wells. Moreover, the similar construction of wells in specific fields increases the chances that, if contamination has occurred, it is already extensive.

Lastly, the need for the "full" casing and cementing of Class II wells is generally less because-brine and other fluids associated with oil and natural gas production pose less threat to human health than fluids which Class I and some Class III wells often inject.

PATHWAY 3 - MIGRATION OF FLUIDS FROM AN INJECTION ZONE THROUGH THE CONFINING STRATA

The third way by which fluids can enter an underground source of drinking water is from an injection zone through the confining strata. Upon entry into an injection zone, fluids injected under pressure will normally travel away from the well laterally and through the receiving formation. In most cases, this expected occurrence gives rise to no concern, but, if the confining stratum which separates the injection zone from an overlying or underlying underground source of drinking water is either fractured or permeable, the fluids can migrate out of the receiving formation and into the protected region.

For obvious reasons, there are no well construction standards which can address this problem of migration of fluids through this pathway. Consequently, the regulations propose two provisions to assure that fluids do not travel this pathway into underground drinking water. First, the regulations require that, prior to the issuance of a permit, the geologic characteristics of the injection zone and confining strata be reviewed.[34] Data already available to states can assist Directors in making these reviews. A permit should only be issued upon the Director's finding that the underground formations are sufficiently sound to contain fluids in the injection zone.

Second, the regulations require that well injection pressure be controlled to prevent opening fractures in the confining strata or otherwise causing the rise of fluids into an overlying protected zone.[35] Using this mechanism, injection pressures can be restricted to provide conservative protection even in the face of less than ideal geologic characteristics. For example, if a confining stratus is known to be fractured or permeable, injection might be permissible if done at predetermined pressure levels which under no circumstances could cause a rise of fluid to the height at which it would enter a drinking water source.

PATHWAY 4 - VERTICAL MIGRATION OF FLUIDS THROUGH IMPROPERLY ABANDONED AND IMPROPERLAY COMPLETED WELLS

One of the common ways by which fluids can enter an underground source of drinking water is by migration through improperly abandoned and improperly completed wells. This would occur if fluids moving laterally within an injection zone encountered an improperly abandoned or completed well, and, following the path of least resistance, flowed upward within the well until entering an overlying underground source of drinking water or overflowing onto the land surface. Because of the large number of wells drilled in the past, and because well operation and abandonment have not always benefitted from close regulatory scrutiny, contamination by this route can present a significant risk to public health. It is estimated that there are about 17,000 improperly abandoned or improperly completed wells which could cause this problem.[36]

To prevent this contamination, the regulations require Directors to determine an "area of review" for injection wells. This is the area around the injection well through which the incremental pressure of injection can cause vertical migration. Operators of Class I, III, and new Class II wells (operators of existing and converted Class II wells are treated differently; see below) must locate other wells within the "area of review" and correct any problems related to improperly abandoned or improperly completed wells before beginning injection.[37] Under this approach, well injectors would have the affirmative responsibility to demonstrate that the proposed injection operation would not cause contamination by this route.

Directors could choose either of two methods to determine the area of review. the first method would be to require use of mathematical formulae to determine, on a case by case basis, the lateral impact which an injection operation could cause. The formula would indicate the distance outward from the well which this particular injection would or could affect. The Regulations provide one formula which can be used for this purpose. It takes into account a range of factors, including hydraulic conductivity, thickness of the injection zone, time of injection, storage coefficient, injection rate, hydrostatic head and specific gravity. EPA is proposing this particular formula because it is based on an equation which has been in common use for years and, in that time has demonstrated satisfactory results; however, other suitable formulae are acceptable.[38]

If a suitable formula indicates that no problem exists, injection could commence without any obligation to repair faulty wells found within the area of review. If it did indicate a problem, however, the well operator would be expected to correct it. Correcting the problem could mean that the well operator would have to plug a faulty well at his/her expense. In other cases, the operator might simply have to modify injection pressure to assure that the rise of fluids caused would not cause fluids to enter an underground source of drinking water.

The use of a formula to determine the area of review may not always be feasible. In some instances, necessary information may be lacking. Such formulae also do not have universal applicability: Mathematical formulae, because they are based on ideal conditions (that aquifers are homogeneous, isotropic, and infinite in extent, for example), may not always reflect actual subsurface conditions. Moreover, they assume radial flow in all directions and, in some cases, will not yield a finite distance measurement for well review purposes.

Because of these possibilities, the regulations offer a second method for determining the area of review. Directors may use (in lieu of a case by case formula) a fixed radius of one-quarter mile or greater. The Agency selected this minimum radius after consideration of current state practices and after applying it to a randomly selected population of well fields representing various geologic conditions. EPA had considered use of more extensive review requirements, particularly the use of one-half mile radius for area of review computation, but decided-against them because the less rigorous requirement is more cost-effective, and the one-fourth mile radius proved satisfactory in actual applications.[39] In many cases, use of a larger fixed radius would result in duplicative review of the same wells.

Moreover, the quarter-mile radius is compatible with coverage practiced in most states. Generally, states impose review requirements on we'll operators in a range of 1000 feet from the injection site up to two miles. EPA's selection of the quarter-mile distance represents its assessment of the effectiveness of these varying requirements in the state programs.[40]

EPA has modified the area of review requirement for Class II wells.[41] Unlike the proposal for wells in Classes I and III, the regulations require that only new Class II wells observe area of review requirements. Class II is characterized by large numbers of wells clustered in oil fields. Because new injection wells are interspersed with existing Class II wells, imposing the area of review requirements on new Class II wells should still result in discovery and correction of all faulty wells within the existing well fields, although over a more extended time frame.[42] The Agency opted for this approach because it deemed it to be effective, both from an environmental and cost perspective, and because it considers placing expenses on new, rather than existing, well operators to be a preferable regulatory approach.[43]

With respect to corrective action itself, the regulations impose a flexible standard. Corrective action required for each well will be fashioned by the Director on a case by case basis after considering a variety of site specific criteria.[44] EPA prefers this approach because of the variety of problems or conditions which can trigger the need for corrective action. In one instance, the only corrective action which may be needed to prevent the migration of fluids into an underground source of drinking water through a faulty well might be a reduction of the pressure at which fluids are injected. In other instances, monitoring of nearby wells coupled with a contingency plan to remedy any problems which result from the injection operation might be feasible. In still other cases, it might be necessary to correct the wells. This range of possibilities, as well as the significant costs which corrective action can generate; have encouraged the Agency to adopt the more flexible approach.

PATHWAY 5 - LATERAL MIGRATION OF FLUIDS FROM WITHIN AN INJECTION ZONE INTO A PROTECTED PORTION OF THAT STRATUM

In the most cases, the injection zone of a particular well will be physically segregated from underground sources of drinking water by impermeable materials. In some instances, however, well injectors, may inject into an unprotected portion of an aquifer which in another area will be designated for drinking water purposes. In this event, there may be no impermeable layer or other-barrier to prevent migration of fluids into underground drinking water.

Injection into unprotected portions of aquifers which contain drinking water in other areas must be done with great care. The regulations control this potentially dangerous activity by according the Director a range of construction and operating controls which can be imposed at his/her discretion.[45] Notwithstanding the discretionary controls afforded the Director, specific information must be considered by him prior to allowing injection into such an aquifer. The Director must consider such factors as the injection pressure, the nature of the fluid injected, specific geologic and hydrogeologic conditions, ground water use patterns and other factors. Usually, Directors can allow injections of this type if the predominant flow of the aquifer is such that injected fluids will tend to move away from, rather than toward, the protected part of the aquifer. Even if that is not the case, however, Directors could still allow the injection if any of a variety of operational conditions were satisfied. For example, the Director might allow an injection upon a determination that the rate of flow or the volume or pressure of injection was sufficiently small to assure that fluids would not enter the prote-cted region.

Pathway 6 - Direct Injection of Fluids into or Above an Underground Source of Drinking Water

The last pathway of contamination of groundwater is potentially the most worrisome. The injection of fluids into or above underground sources of drinking water can present the most immediate risk to public health because it can directly degrade groundwater especially if the injected fluids do not benefit from any natural attenuation from contact with soil, as they might during movement through an aquifer or separating stratum.

The regulations prohibit injection of contaminants directly into an underground source of drinking water for wells in Classes I to III;[46] Class IV wells, which inject directly into underground drinking water are to be banned as soon as possible but in no event later than six months after a State underground injection control program becomes effective. Class IV wells which inject above an underground drinking water source are to be studied further. Accordingly, EPA has decided to defer issuance of permitting and technical standards for Class IV wells until this fall. Class V wells, of which little is known, will be assessed before regulations for their operation are proposed[47] (for a fuller discussion of the regulatory approach proposed for Class IV and V wells, see the preamble to the regulations).

Other Requirements

Abandonment - the regulations also require that well injectors abandon their injection wells in a way which will prevent the contamination of underground sources of drinking water.[48], [49] As indicated earlier, abandoned wells can act as conduits for contaminants to enter protected aquifers. To assure that currently used and future wells do not create problems of this type, the regulations require plugging of wells after termination of operation. Again, the exact means of accomplishing an effective abandonment are left to the judgment of the Director to be exercised on a case by case basis. In addition, §146.10(d) requires the operator of a Class III well to subinit a plan of abandonment which must demonstrate that no movement from the mining zone into underground sources of drinking water will occur after abandonment.

§146.10(b) specifies that the dump bailer method, the balance method, or the two plug method be used to plug a well. These practices represent existing methods employed that have proved effective in a wide range of application.[50]

With respect to Class IV wells, traditional methods of abandonment, such as plugging, may be inappropriate due to the crude construction of the well. In such a case, the only abandonment requirements might simply be closure of the well.

MONITORING FREQUENCIES

The regulations also require various kinds of monitoring.[51] Monitoring can provide an early warning of potential serious degradation of underground sources of drinking water.

Wells in Classes I, II, and III share common monitoring requirements. Injection fluids must be tested with sufficient frequency to field data representative of fluid characteristics. Information of this sort is essential for the Director to understand the operation of a particular well. Such information serves the important function of providing basic knowledge of enabling Directors to analyze reasons for well failures, to establish appropriate remedies to correct them and to assess any endangerment the failures might cause.

The regulations also require monitoring of operating characteristics of wells in Classes I, II, and III. Class I and III wells must have continuous recording devices to monitor injection pressure, flow rate, and volume of injection fluids.[52] Continuous monitoring is appropriate because fluids injected by Class I and III wells are usually more corrosive and hazardous than are fluids injected by others. These fluid properties increase the risk of serious well leaks or failures. Continuous monitoring, furthermore, is a common practice for these wells, in part because they often inject fluids in uninterrupted streams.

Class I wells must comply with the additional requirement of continuously monitoring the pressure in the annulus of the well between the tubing and the long string. The "long string" is the casing which extends from the ground surface to the injection zone. Wells in Class III which may require the use of the annulus for injection, need not meet this requirement since, when the annulus is employed for injection, pressure measurements reflect injection pressure.

Class II injection well monitoring provisions aze less stringent than those for Classes I and III.[53] Continuous monitoring is not required for Class II; rather, depending on the actual injection operation, monitoring frequency varies from daily to monthly. A stricter approach is not essential for Class II wells because of the lesser toxicity and corrosivity of fluids which Class II wells handle and because the total cost of imposing continuous monitoring on Class II wells would have been inordinately burdensome in EPA's view.[54]

Class III wells are also required to monitor, on a quarterly basis, water supply wells adjacent to the injection site to detect any excursions from the injection site.[55] This monitoring is commonly practiced by operators of Class III wells.[56] EPA is promulgating this requirement for Class III wells (and not for Class I wells) because Class III wells are often designed to inject into shallower strata, thereby increasing the possibility of contamination of aquifers nearest the land surface.

This added risk has prompted the Agency to require monitoring wells at each project site, located to maximize the probability of detecting any horizontal or vertical fluid excursion from the injection zone. Weekly monitoring of the fluid levels in these monitoring wells and of parameters appropriate to determine if any excursions of injected fluids are entering underground sources of drinking water, is also required. This requirement, although involving additional expense, was considered necessary to assure that any migration of these potentially harmful injected fluids into underground sources of drinking water, which are often located quite close to the injection zones, would be discovered and rectified promptly. Class III wells may be monitored on a field or project basis rather than an individual well basis by manifold monitoring e.g., using a common header with individual well points. This approach may be used with facilities that consist of more than one injection well if the owner or operator can demonstrate that manifold monitoring is comparable to individual well monitoring.[57]

No monitoring requirements are proposed for Class V wells. These wells will be assessed under the proposed regulatory scheme. The assessment should produce a substantial amount of data upon which an entire regulatory approach, including monitoring, can be used.

Reporting

The regulations also impose reporting requirement on well injectors.[58] Owners and operators of wells regulated under Classes I and III must report the results of monitoring and any other significant operational information at least quarterly, while Class II well owners and operators need only report to the director annually. The reasons underlying these proposals parallel those for the monitoring requirements. Owners and operators of wells which inject fluids of greater potential hazard must report more often than those which do not. Class V wells need not submit monitoring or reporting data because the assessment planned for this category will supply EPA with a substantial amount of data in its own right.

End Notes

[1] Interviews with State officials from Texas, Florida, Kansas, California and Michigan.

[2] Preliminary Injection Well Practices, Geraghty and Miller, Inc., Tampa, Florida; 1977.

[3] See Generally The Report to Congress, Waste Disposal Practices and Their Effects on Ground Water, U.S. Environmental Protection Agency (January, 1977), Sections V, VIII, XIII ("Report to Congress"); A Manual of Laws, Regulations, and Institutions-for Control of Groundwater Pollution, U.S. Environmental Protection Agency, (June, 1976), Chapter I ("Manual"). Underground Injection Control Regulations, Subpart F, Injection Well Practices, Draft Final Report, Geraghty and Miller, Inc. and Temple, Barker and Sloane, Inc. March, 1978 ("Subpart F"). Preliminary Evaluation of- Well Injection Practices, Geraghty and Miller, ("Preliminary Evaluation").

[4] §§ 146.12(b); 146.22(b); 146.32(a).

[5] See Generally Report to Congress, Section XI, XIII; Manual, Chapter IC: Preliminary Evaluation; Ground Water Pollution From Subsurface Excavations, U.S. Environmental Protection Agency, 1973, Part 2, Section II, Ground Water Pollution."

[6] Report to Congress, Sections III, IV; Manual, Chapter I, pp. I-10 -- 1-50.

[7] Report to Congress, Section VIII, IX, XIII: Manual, Chapter I; Subpart F; Preliminary Evaluation.

[8] Comprehensive Radioactive Waste Management Program Message from the President Received during Recess, Congressional Record - Senate, February 18, 1980.

[9] Development of Procedures for Subclassification of Class III Injection Wells, Geraghty and Miller, April 30, 1980.

[10] Report to Congress, Section XI, XIII; An Introduction to the Technology of Subsurface Wastewater Injection, U.S. Environmental Protection Agency (December, 1977), Chapter 7 ("Subsurface Wastewater Injection").
[11] §146.12(b). For a full discussion of the requirements for Class II, and III wells, see pages 11-12 below.
[12] Subsurface Wastewater Injection, Chapter 7.
[13] §146.08.
[14] See generally, Mechanical Integrity Testing of Injection Wells, Geraghty and Miller, Inc., April 30, 1980.
[15] §146.08(b),(c).
[16] see generally, Mechanical Integrity Testing of Injection Wells.
[17] §146.08(d).
[18] §§146.13(b),(3), 146.23(b),(3), 146.33(b),(3).
[19] Subsurface Wastewater Injection, Chapter 7.
[20] §146.12(c).
[21] Report to Congress, Section XIII; Ground-Water Pollution, Part 2, Section II.
[22] Subsurface Wastewater Injection, Chapter 7.
[23] §146.12(c)(1).
[24] Subsurface Wastewater Injection, Chapter 7; See also Cook, T.D. Underground Waste Management and Environmental Implications. American Association of Petroleum Geologists (Tulsa, Okla., 1972).
[25] §146.12(c).
[26] §146.08.
[27] §§146.12(b); 146.32(a).
[28] §§ 146.12(b),(1)-(7); 146.22(b),(1)-(7); 146.32(a),(1)-(7).
[29] §146.22(b).
[30] Estimated after discussions with consultants. See generally Cost of Compliance Proposed Underground Injection Control Programs, Oil and Gas Wells, Arthur D. Little, Inc. (June, 1979) ("Oil and Gas Wells").
[31] See generally Underground Injection Control Program Class II Well Incremental Compliance Cost Refinements. Booz, Allen and Hamilton Inc., and Geraghty and Miller, Inc., April 30, 1980.
[32] Report to Congress, Section XI.
[33] Id.
[34] §§146.14(a)(1); 146.24(a)(1); 146.34(a)(1).
[35] § 146.06.
[36] Oil and Gas Wells; Chapter VIII-D.
[37] §122.44(a).
[38] §146.06(a).
[39] See generally Oil and Gas wells, Chapter VIII.
[40] Preliminary Evaluation of Well Injection Practices.
[41] §146.24(a),(2).
[42] See generally Oil and Gas Wells, Chapter VIII.
[43] Id.
[44] §§146.07, 146.14.
[45] §146.12; 146.22; 146.32.
[46] §122.34(a)(1).
[47] §146.52(b)(1).

[48] §146.10.

[49] See Generally, Development of Procedures and Costs for Proper Abandonment and Plugging of Injection Wells, Booz Allen and Hamilton Inc., and Geraghty and Miller, April 30, 1980.

[50] Development of Procedures and Costs for Proper Abandonment and Plugging of Injection Wells.

[51] §§146.13; 146.23; and 146.33.

[52] §§146.13(b)(2); and 146.33(b)(2).

[53] §146.23(b).

[54] Oil and Gas Wells, Chapter V-B, C.

[55] §146.33(b)(5).

[56] Comments of Freeport Sulfur Co., Jan. 14, 1977; Statement by Texas Gulf Co., Oct. 13, 1976.

[57] §146.33(b)(6).

[58] §§ 146.13; 146.23; and 146.33.

INDEX

D

E

F

G

H

I

L

M

N

O

P

Q

R

S

T